Advances in Optical Form and Coordinate Metrology

IOP Series in Emerging Technologies in Optics and Photonics

Series Editor

R Barry Johnson a Senior Research Professor at Alabama A&M University, has been involved for over 50 years in lens design, optical systems design, electro-optical systems engineering, and photonics. He has been a faculty member at three academic institutions engaged in optics education and research, employed by a number of companies, and provided consulting services.

Dr Johnson is an IOP Fellow, SPIE Fellow and Life Member, OSA Fellow, and was the 1987 President of SPIE. He serves on the editorial board of Infrared Physics & Technology and Advances in Optical Technologies. Dr Johnson has been awarded many patents, has published numerous papers and several books and book chapters, and was awarded the 2012 OSA/SPIE Joseph W Goodman Book Writing Award for Lens Design Fundamentals, Second Edition. He is a perennial co-chair of the annual SPIE Current Developments in Lens Design and Optical Engineering Conference.

Foreword

Until the 1960s, the field of optics was primarily concentrated in the classical areas of photography, cameras, binoculars, telescopes, spectrometers, colorimeters, radiometers, etc. In the late 1960s, optics began to blossom with the advent of new types of infrared detectors, liquid crystal displays (LCD), light emitting diodes (LED), charge coupled devices (CCD), lasers, holography, fiber optics, new optical materials, advances in optical and mechanical fabrication, new optical design programs, and many more technologies. With the development of the LED, LCD, CCD and other electo-optical devices, the term 'photonics' came into vogue in the 1980s to describe the science of using light in development of new technologies and the performance of a myriad of applications. Today, optics and photonics are truly pervasive throughout society and new technologies are continuing to emerge. The objective of this series is to provide students, researchers, and those who enjoy self-teaching with a wide-ranging collection of books that each focus on a relevant topic in technologies and application of optics and photonics. These books will provide knowledge to prepare the reader to be better able to participate in these exciting areas now and in the future. The title of this series is Emerging Technologies in Optics and Photonics where 'emerging' is taken to mean 'coming into existence,' 'coming into maturity,' and 'coming into prominence.' IOP Publishing and I hope that you find this Series of significant value to you and your career.

Advances in Optical Form and Coordinate Metrology

Edited by
Richard Leach
Manufacturing Metrology Team, Faculty of Engineering, University of Nottingham, Nottingham, UK

IOP Publishing, Bristol, UK

ISBN 978-0-7503-2524-0 (ebook)
ISBN 978-0-7503-2522-6 (print)
ISBN 978-0-7503-2525-7 (myPrint)
ISBN 978-0-7503-2523-3 (mobi)

DOI 10.1088/978-0-7503-2524-0

Version: 20201201

IOP ebooks

British Library Cataloguing-in-Publication Data: A catalogue record for this book is available from the British Library.

Published by IOP Publishing, wholly owned by The Institute of Physics, London

IOP Publishing, Temple Circus, Temple Way, Bristol, BS1 6HG, UK

US Office: IOP Publishing, Inc., 190 North Independence Mall West, Suite 601, Philadelphia, PA 19106, USA

Contents

Preface x

Editor biography xi

List of contributors xii

1 Terms, definitions and standards **1-1**
Richard Leach

1.1 Introduction 1-1
1.2 Surface and coordinate terms and definitions 1-2
1.3 General metrology terms and definitions 1-3
1.4 Specification standards for coordinate metrology 1-6
1.4.1 Coordinate metrology standards 1-6
Acknowledgements 1-7
References 1-7

2 State-of-the-art in point cloud analysis **2-1**
Sofia Catalucci and Nicola Senin

2.1 Introduction 2-1
2.1.1 Mathematical representation of 3D point clouds 2-2
2.1.2 Common 3D point cloud file formats 2-2
2.1.3 Point cloud transformations 2-3
2.2 Extracting properties and organising information in point clouds 2-7
2.2.1 Convex hull and bounding boxes 2-7
2.2.2 Centroid and principal axes of a point cloud 2-8
2.2.3 Spatial subdivision of point clouds 2-10
2.3 Point cloud pre-processing 2-14
2.3.1 Point cloud simplification, decimation and resampling 2-15
2.3.2 Elimination of isolated points and noise reduction 2-16
2.3.3 Triangle meshes and conversion to/from point clouds 2-18
2.4 Point features and partitioning 2-20
2.4.1 Point normals 2-20
2.4.2 Point curvatures 2-21
2.4.3 Partitioning and segmentation 2-23
2.5 Point cloud fitting 2-25
2.5.1 Fitting methods 2-25

2.6 Registration of point clouds 2-29
2.6.1 Registration based on external references or based on matching landmarks 2-32
2.6.2 The absolute orientation problem 2-32
2.6.3 Alignment by means of principal component analysis 2-33
2.6.4 RANSAC alignment 2-34
2.6.5 Alignment by iterative closest points 2-36
2.6.6 Landmark matching and alignment using similarity metrics 2-38
2.7 Measurement uncertainty in point cloud surface data 2-38
2.7.1 Approaches to the estimation of measurement uncertainty 2-39
2.7.2 Uncertainty associated with point clouds 2-41
2.8 Conclusions 2-42
References 2-43

3 Laser triangulation 3-1
Mohammed A Isa, Samanta Piano and Richard Leach

3.1 Laser triangulation 3-1
3.2 Laser triangulation sensors 3-3
3.3 Laser triangulation measurement dependence on surface properties 3-5
3.3.1 Measurement uncertainty limit 3-6
3.3.2 Surface reflectance perspective 3-7
3.3.3 Measurement dependence on surface form 3-11
3.4 Laser triangulation systems 3-13
3.4.1 Extension of a point based laser triangulation sensor 3-13
3.4.2 Point measurement systems 3-15
3.4.3 Profile measurement systems 3-17
3.4.4 Surface measurement systems 3-19
3.4.5 Advanced laser triangulation systems 3-21
3.5 Working process of laser triangulation 3-24
3.5.1 Characterisation of intrinsic parameters 3-24
3.5.2 Characterisation of extrinsic parameters 3-25
3.5.3 Pre-calibration of laser triangulation systems 3-25
3.5.4 Scanning path planning 3-27
3.5.5 Image pre-processing 3-27
3.5.6 Laser feature extraction 3-28
3.5.7 Refinement and postprocessing 3-30

3.6 Application of laser triangulation measurements 3-31
3.6.1 Application of laser triangulation in emerging manufacturing methods 3-31
3.6.2 Application for geometric inspection 3-33
3.6.3 Application of 3D reconstructed models 3-34
3.7 Conclusions 3-35
References 3-36

4 Close-range photogrammetry 4-1
Danny Sims-Waterhouse, Richard Leach and Samanta Piano

4.1 Introduction 4-1
4.1.1 Modern photogrammetry 4-2
4.1.2 Camera projection theory 4-2
4.1.3 The pinhole camera model 4-3
4.1.4 Distortion modelling 4-5
4.2 Characterisation and calibration 4-6
4.2.1 Camera characterisation 4-6
4.2.2 Linear techniques 4-7
4.2.3 Non-linear methods 4-8
4.2.4 Self-calibration 4-11
4.3 System calibration 4-11
4.3.1 Arbitrary scale 4-11
4.4 Image acquisition 4-12
4.4.1 Depth of field 4-12
4.4.2 Imaging procedure optimisation 4-14
4.5 Summary 4-16
References 4-16

5 Digital fringe projection profilometry 5-1
Rui Chen, Jing Xu and Song Zhang

5.1 Introduction 5-1
5.2 Fringe pattern generation methods 5-1
5.2.1 Conventional fringe generation methods 5-1
5.2.2 Digital binary defocusing techniques 5-2
5.2.3 New fringe pattern generation methods 5-3
5.3 Fringe analysis 5-4
5.3.1 Conventional fringe analysis techniques 5-4
5.3.2 Machine learning enhanced fringe analysis methods 5-6

5.4 Phase unwrapping 5-6
5.4.1 Conventional phase unwrapping methods 5-7
5.4.2 Non-conventional phase unwrapping methods 5-8
5.4.3 Machine-learning-based phase unwrapping methods 5-9
5.5 High dynamic range techniques 5-9
5.6 Calibration 5-11
5.6.1 Conventional methods 5-11
5.6.2 Recent developments 5-12
5.7 High-speed FPP realisation 5-13
5.7.1 Conventional high-speed FPP methods 5-13
5.7.2 Recent developments 5-14
5.8 Towards automation 5-15
5.9 Towards integrated solutions 5-16
5.9.1 3D imaging system for crime scene evidence collection 5-17
5.9.2 Robotic path planning 5-17
5.10 Summary 5-19
References 5-19

6 Machine learning approaches **6-1**
Joe Eastwood, Danny Sims-Waterhouse and Samanta Piano

6.1 Introduction 6-1
6.2 Overview of machine learning and machine learning methods 6-2
6.3 Machine learning for stereo matching 6-7
6.3.1 Learned stereo machines 6-9
6.4 Machine learning for phase unwrapping 6-10
6.5 Learning depth from a single image 6-12
6.5.1 Characterisation of cameras and projectors 6-13
6.6 Machine learning for point cloud analysis 6-14
6.6.1 Point cloud segmentation 6-15
6.6.2 Point cloud registration 6-18
6.6.3 Point cloud completion 6-20
6.7 Conclusions 6-21
References 6-21

7 Precision freeform metrology **7-1**
Chi Fai Cheung, Lingbao Kong and Mingjun Ren

7.1 Overview of freeform surfaces 7-1
7.2 Framework for precision freeform metrology 7-3

7.3 Characterisation of freeform surfaces 7-6
 7.3.1 Surface fitting and reconstruction 7-6
 7.3.2 Surface matching 7-10
 7.3.3 Surface parameters for form error evaluation 7-14
7.4 Conclusions and future research 7-15
 Acknowledgments 7-16
 References 7-16

8 Performance verification for optical co-ordinate metrology 8-1
Adam Thompson and Nicholas Southon

8.1 Introduction 8-1
8.2 Material measures 8-2
 8.2.1 Material measure 1 8-2
 8.2.2 Material measure 2 8-3
 8.2.3 Material measure 3 8-4
 8.2.4 Material measure 4 8-4
8.3 The different types of performance verification test 8-5
8.4 Specification of errors 8-6
 8.4.1 General guidelines for performance verification 8-7
8.5 Performance verification procedures 8-8
 8.5.1 Measurements with representative points 8-9
 8.5.2 General preparation for the performance verification test 8-10
 8.5.3 Probing characteristics 8-11
 8.5.4 Distortion characteristics 8-13
 8.5.5 Length measurement errors 8-20
8.6 Compliance with specifications 8-23
 8.6.1 Formal definition of performance verification 8-23
 8.6.2 Retesting after a failed compliance with specifications 8-24
8.7 A final note on the validity of performance verification 8-24
 References 8-25

Preface

The book will cover the latest advances in the development of optical form and coordinate metrology instruments and the recent ISO framework on their performance verification. Latest advances discussed will include the drive towards faster instruments for in-process applications, the ability to measure highly complex shapes (in, for example, additive manufacturing), advances in the use of artificial intelligence to enhance data analysis and the push towards calibration.

Surface form and coordinate measurement is essential and ubiquitous in many industries, from manufacturing to automotive and aereospace and cultural heritage applications. However, in this book we focus our dialogue and applications on manufacturing. The dimensions of an object are often the most critical features when considering manufacturing tolerances, assembly with other parts and ultimately the functionality of the object. In addition, the range and complexity of shapes being manufactured has recently increased significantly as the adoption of additive manufacturing is embraced more widely. Shape measurement in manufacturing is often carried out using contact probing, which is slow, requires contact, and produces small data sets (or minimal surface coverage). To increase throughput and allow the adoption of digital approaches, industry needs to use faster optical (non-contact) methods that produce dense point clouds. However, optical methods can be labour-intensive and expensive, especially when high-accuracy dense object coverage is required. There are a large number of available optical instruments for form and coordinate metrology. The book will highlight the latest advances in the instrumentation and calibration of such instruments, therefore, informing users and instrument developers of the latest techniques on offer or around the corner.

Editor biography

Richard Leach

Richard is currently a Professor in Metrology at the University of Nottingham and prior to this spent 25 years at the National Physical Laboratory. He has been researching and lecturing on surface metrology for over 30 years. He is on the Council of the European Society of Precision Engineering and Nanotechnology, the Board of the American Society of Precision Engineering and several international standards committees. He is the European Editor-in-Chief for Precision Engineering and has over 500 publications including eight textbooks. He is a Fellow of the Institute of Physics, the Institution of Engineering & Technology, Higher Education Authority, the Institute of Measurement & Control, the International Society of Nanomanufacturing and the International Academy of Production Engineering (CIRP). He is a visiting professor at Loughborough University and the Harbin Institute of Technology.

List of contributors

Sofia Catalucci

University of Nottingham, Nottingham, UK

Sofia Catalucci is a researcher at the University of Nottingham within the Manufacturing Metrology Team. Her research is in the development of smart, adaptive measurement systems, capable of automated planning and assessment of measurement quality, for application in flexible manufacturing environments, in particular involving additive manufacturing. Currently, she is focusing on knowledge-driven algorithmic point cloud processing for optical instruments, applied to complex freeform hollow components, and the development of quality metrics and uncertainty indicators on point clouds.

Rui Chen

Tsinghua University, Beijing, People's Republic of China

Rui Chen is a researcher in Tsinghua University, Beijing, China in the Department of Mechanical Engineering. His research interests include three-dimensional measurement and cyber physical systems.

Chi Fai Cheung

The Hong Kong Polytechnic University, Kowloon, Hong Kong

Benny Cheung is Professor and the Head of the State Key Laboratory of Ultra-precision Machining Technology at the Department of Industrial and Systems Engineering of The Hong Kong Polytechnic University. He is an Adjunct Professor of Changchun University of Science and Technology. His primary research interest is in precision engineering, including ultra-precision machining technology, precision metrology and precision manufacturing. His research in freeform metrology covers a wide spectrum of activities encompassing instrumentation for freeform surface measurement, surface modelling and reconstruction, surface characterisation, multi-sensor data fusion and uncertainty analysis.

Joe Eastwood

University of Nottingham, Nottingham, UK

Joe Eastwood is a researcher in the Manufacturing Metrology Team at the University of Nottingham. His research focuses on the development of a fully automated measurement pipeline for close-range photogrammetry via the exploitation of *a priori* data through machine learning.

Mohammed A Isa

University of Nottingham, Nottingham, UK

Mohammed A Isa is a research fellow at the University of Nottingham within the Manufacturing Metrology Team. Prior to this position, he worked on multi-axis additive manufacturing and 3D reconstruction of freeform surfaces. Currently, his research interests in measurement systems include laser triangulation, photogrammetry and precision motion systems. His research in optical 3D coordinate metrology uses machine vision, multi-view geometry and intelligent error modelling to improve measurement uncertainty.

Lingbao Kong
Fudan University, Shanghai, People's Republic of China
Bobby Kong is a Research Professor and Deputy Director of Shanghai Engineering Research Centre of Ultra-precision Optical Manufacturing, Fudan University, China since 2015. He joined The Hong Kong Polytechnic University as a Research Assistant in 2004, a Research Associate in 2008 and then a Scientific Officer in 2011. His research interests include ultra-precision optical manufacturing, freeform measurement and characterisation, intelligent machine vision and precision measurement, multi-sensor measurement, uncertainty in optical measurement, design and fabrication of functional structures, and the development of precision equipment.

Samanta Piano
University of Nottingham, Nottingham, UK
Samanta Piano is an Assistant Professor at the University of Nottingham within the Manufacturing Metrology Team. Her current research interest concerns the development of innovative optical techniques and 3D probing systems for high-precision coordinate metrology to be used in industrial applications. Her broad expertise includes surface analysis, scanning probe spectroscopy, atomic magnetometry and quantum sensing.

Mingjun Ren
Shanghai Jiao Tong University, Shanghai, People's Republic of China
Mingyu Ren is an Associate Professor at Shanghai Jiao Tong University, China. His current research interests include ultra-precision manufacturing and *in situ* measurement and intelligent instrumentation. He is currently on the Youth Committee of China Instrument and Control Society, the Standing Committee of Division of Measurement and Control, and the China Instrument and Control Society.

Nicola Senin
University of Perugia, Perugia, Italy
Nicola Senin is an Associate Professor in Manufacturing Technologies and Systems at the Department of Engineering, University of Perugia, Italy, and a visiting researcher in the Manufacturing Metrology Team at the University of Nottingham. His primary research interests are in coordinate metrology and in the area of in-process sensing, monitoring and control for advanced manufacturing processes. He is member of international scientific and technical societies in manufacturing and precision engineering.

Danny Sims-Waterhouse
Taraz Metrology, UK
Danny Sims-Waterhouse is a Development Engineer at Taraz Metrology, focussing on the development of optical measurement systems for the additive

manufacturing industry. The development work aims to provide metrology solutions at all stages of the manufacturing process, including in-line and post-process measurement systems. Prior to his position, he carried out research within the Manufacturing Metrology Team at the University of Nottingham. His research is mainly focussed on the development of photogrammetry methods for high accuracy form measurement, fringe projection, machine learning and smart algorithm development.

Nicholas Southon
INSPHERE Ltd, Bristol, UK
Nicholas Southon is a metrology engineer with INSPHERE, developing innovative metrology solutions for the energy, automotive and aerospace sectors. Previously, he was a research associate with the University of Nottingham in the Manufacturing Metrology Team. His research is focussed on in-process metrology and performance verification for optical metrology.

Rong Su
University of Nottingham, Nottingham, UK
Rong Su joined the University of Nottingham as a Research Fellow within the Manufacturing Metrology Team in 2015 following a period working at the National Physical Laboratory in the UK. His expertise includes the theoretical modelling, calibration and development of 3D imaging and light scattering instruments for surface and dimensional metrology. He is a member of the editorial board of *Light: Advanced Manufacturing* and *Nanomanufacturing and Metrology*, and a scientific committee member of several professional societies.

Adam Thompson
University of Nottingham, Nottingham, UK
Adam Thompson is a Research Fellow at the University of Nottingham in the Manufacturing Metrology Team. Adam's research background is in surface topography measurement of additively manufactured parts, particularly in the measurement of metal and polymer parts using X-ray computed tomography. Adam also has expertise across other areas of metrology, having undertaken research projects in performance verification of fringe projection and X-ray computed tomography systems. Adam has a deep understanding of coordinate and in-process measurement principles, having taught numerous undergraduate and postgraduate courses in basic advanced metrology and performed research activities in the area. Prior to his PhD, Adam taught high school physics.

Jing Xu
Tsinghua University, Beijing, People's Republic of China
Jing Xu was a Postdoctoral Researcher in the Department of Electrical and Computer Engineering, Michigan State University, East Lansing, USA, and is currently an Associate Professor in the Department of Mechanical Engineering, Tsinghua University, Beijing, China. His research interests include vision-guided manufacturing, image processing and intelligent robotics. He served as an

associate vice president of the IEEE Robotics and Automation Society from 2019 to 2020.

Song Zhang
Purdue University, West Lafayette, IN, USA
Song Zhang is Professor in Mechanical Engineering and the Assistant Head for Experiential Learning in the School of Mechanical Engineering at Purdue University, USA. He previously worked at Iowa State University, USA, as a professor and at Harvard University, USA, as a postdoctoral research fellow. His primary research area is the investigation of various fringe projection techniques for high-speed 3D optical metrology, the development of novel algorithms to compress 3D data for efficient 3D video communication and data storage, and the exploration of novel applications in other disciplines through collaborations. He has served as Chair for many conference symposiums, and as an associate editor for *Optics and Lasers in Engineering* and *Optics Express*.

IOP Publishing

Advances in Optical Form and Coordinate Metrology

Richard Leach

Chapter 1

Terms, definitions and standards

Richard Leach

In this chapter, some common terms and definitions are presented and discussed. Terms and definitions relating to metrology in general are given along with those specific to coordinate metrology. Throughout the chapter, the relevant international standards are referred to and referenced.

1.1 Introduction

Surface shape (or form) measurement is essential and ubiquitous in many industries, including diverse areas such as cultural heritage applications, but this book will focus on applications in engineering and manufacturing. The dimensions of an object are often the most critical features when considering engineering tolerances, the assembly with other parts and ultimately the functionality of the object (Schmitt *et al* 2016, Hocken and Pereira 2016). In addition, the range and complexity of shapes being manufactured has recently increased significantly as the adoption of additive manufacturing is embraced (Stavroulakis and Leach 2016). Shape measurement in manufacturing is often carried out using contact probing, which is slow, requires contact and produces small datasets (or minimal surface coverage). To increase throughput and allow the adoption of digital approaches, industry wishes to use faster optical (non-contact) methods that produce dense point clouds. Cost-effective optical methods have recently become available, and this book is designed to summarise the recent advances in their design, operation and calibration.

This chapter presents and discusses some common terms and definitions used throughout the book. Chapter 2 summarises the important field of point cloud analysis, which is critical for all optical form and coordinate measuring instruments. Chapters 3 through 5 present the latest advances in laser triangulation, photogrammetry and fringe projection, respectively. Chapter 6 reviews the latest advances in the use of machine learning for optical measurement and point cloud analysis. Chapter 7 reviews methods for performance verification of optical instruments.

doi:10.1088/978-0-7503-2524-0ch1

1.2 Surface and coordinate terms and definitions

A coordinate measuring system (CMS) is an instrument for measuring the physical geometry of an object and coordinate metrology is the science and application of such measurements. Typically, a CMS has the ability to move a probing system (sensor) that detects the location of the surface on an object and the capability to determine spatial coordinate values on the surface relative to a reference coordinate system.

The following definitions are used in this book:

Surface form—the underlying shape of a part (Leach 2014) or fit to a measured surface (ISO 10110 part 8 (ISO 2010a)).

Surface topography—all the surface features treated as a continuum of spatial wavelengths (Leach 2014). Basically, here surface topography is everything that makes up the geometry of the object's surface, i.e. it is the surface form plus the surface texture.

Surface texture—the geometrical irregularities present at a surface. Surface texture does not include those geometrical irregularities contributing to the form or shape of the surface (Leach 2015). Simply put: surface texture is what is left of the surface topography once the surface form has been removed. Figure 1.1 illustrates the relationship between surface topography, form and texture.

Form and texture make up topography, but it is not always obvious where to draw the distinction between them. Many modern CMSs are equipped with sensors for measuring both form and texture. Some modern surfaces, for example from additive manufacturing (Leach *et al* 2019) can also blur the lines between what is considered texture and what is considered form, as the fine-scale structure is often of the same order in size as the features that make up the shape of the object. Despite this blurred distinction, coordinate and surface metrology tend to be very different fields, including different specification standards and ways in which measurement traceability is demonstrated.

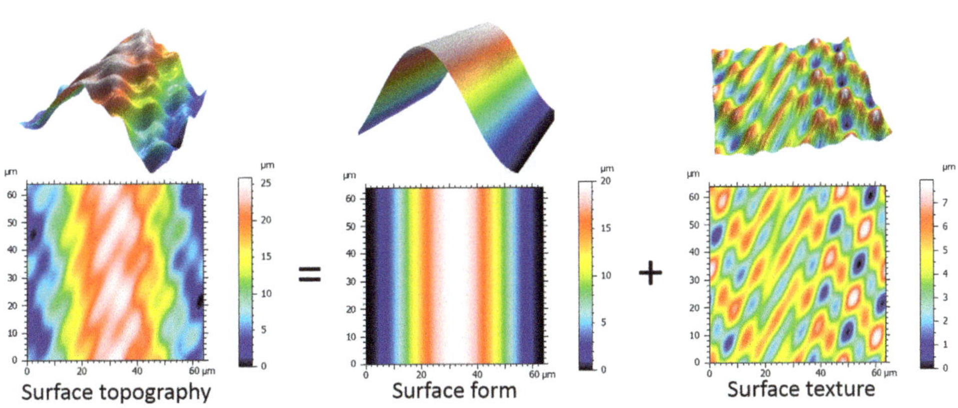

Figure 1.1. Different surface types.

The geometrical form of a part can be critical for its effective implementation. Form is mathematically described as the deviation of the measured surface from the fundamental definition of the corresponding geometrical primitive, also known as the reference geometry feature. The following form parameters are common to all geometries presented here.

1. *Peak-to-reference deviation*: maximum deviation of measured points in the positive direction with respect to the reference geometry.
2. *Reference-to-valley deviation*: maximum deviation of measured points in the negative direction with respect to the reference geometry.
3. *Peak-to-valley deviation*: a measure of the maximum variation in measured points, distance between the parameters in 1 and 2.
4. *Root-mean-square deviation*: a measure of the distribution of measured points with respect to the reference geometry.

Form deviations are often quoted only as positive values since they are defined for a specific direction from the reference geometry. In some cases, reference-to-valley deviations are quoted as negative values, in which case its absolute value is used in determining peak-to-valley deviation. In this chapter, parallelism is the only feature for which the measurement is a comparison to a physical reference instead of a fundamental geometry. Consequently, the positivity of the deviation is important for defining parallelism.

1.3 General metrology terms and definitions

(Adapted from Leach *et al* 2015.)

There are a number of terms relating to the field of metrology that need to be discussed briefly. Any of these terms are used almost indistinguishably in practice, which can often lead to confusion when specifying instruments. The terms used are taken from the most recent version of the *BIPM International Vocabulary of Metrology* (BIPM *et al* 2012).

Traceability—the concept of traceability is one of the most fundamental in metrology and is the basis upon which all measurements can be claimed to be accurate. Traceability is defined as follows:

> Traceability is the property of the result of a measurement whereby it can be related to stated references, usually national or international standards, through a documented unbroken chain of comparisons all having stated uncertainties.

It is important to note the last part of the definition of traceability, which states *all having stated uncertainties*. This is an essential part of traceability as it is impossible to usefully compare, and hence calibrate, instruments without a statement of uncertainty. Uncertainty and traceability are inseparable (Haitjema 2018).

Traceability applied to optical coordinate metrology is discussed in Leach and Shaheen (2020).

Calibration is defined as follows:

> Operation that, under specified conditions, in a first step establishes a relation between the quantity values with measurement uncertainties provided by measurement standards and corresponding indications with associated measurement uncertainties and, in a second step, uses this information to establish a relation for obtaining a measurement result from an indication.

In simpler terms, calibration is a comparison between two measurements; one of which is a reference or standard value, and the other which is being tested. Again, note the use of the term uncertainty in the formal definition of calibration. Commonly the term calibration is misused, which has led to confusion in understanding the aim of the calibration process. The frequent misuse of the term calibration is when it is confused with adjustment.

Adjustment is defined as follows:

> A set of operations carried out on a measuring system so that it provides prescribed indications corresponding to given values of a quantity to be measured

The adjustment process physically changes some parameters of a metrological tool (it can be a mechanical adjustment or it could be the result of changing the value of a software constant) to provide an indication that is closer to a known value. The adjustment process does not provide information about measurement uncertainty. Similar results could be obtained by correcting the measurement result using the results from a calibration certificate. A meaningful measurement result can be presented without adjustment, but it must have an associated uncertainty.

An example of adjustment of a CMS is when the manufacturer adjusts the coordinate data based on a map of the degrees of freedom of the machine axes (known as software error correction). All modern tactile CMSs (and those with multiple sensors) have software error correction applied, but it is not so popular for optical CMSs. The adjustment cannot account for the uncertainty associated with the measurement result; it only uses a value from the range of possible values that are within the limits given by the measurement uncertainty. After this adjustment, the measurement of an object will provide a different result. The basic difference between calibration and adjustment is also illustrated by the requirement in ISO 17025 (ISO 2005) that an instrument should be calibrated before and after adjustment.

Verification is defined as follows:

> Provision of objective evidence that a given item fulfils specified requirements.

A verification test is designed to check whether a particular instrument attribute meets its specification. Verification, therefore, does not necessarily imply that measurement uncertainty is part of the test, but usually some form of quantitative measure is required. Often, an assertion that an instrument is within specification assumes that the test result is inside the specification by at least a 'guard band', for example, the expanded uncertainty. Verification in optical coordinate metrology is discussed in chapter 8.

Maximum permissible error is a criterion mainly used to quantify the performance of a CMS is the maximum permissible error (MPE), which is defined (in the ISO 10360 standards) as the largest error or deviation of a measurement from a reference quantity value. The MPE of a CMS is often specified as a constant value plus a length-dependent term, for example MPE = $8 + 2L/1000$ μm, where L is the measured length in millimetres. A CMS's MPE serves as a benchmark for its performance when compared to other similar instruments. Often, prior to purchasing a measuring instrument, several tests with dedicated reference objects are performed to determine whether the particular instrument satisfies its specified MPE.

An instrument's MPE is often mistaken for the uncertainty of measurements performed on the instrument. While MPE is a maximum expected deviation of measurements by a particular instrument from a reference value, uncertainty is a statistical dispersion of values that can be attributed to the result of a measurement. A measurement result expressed as a single value corresponding to the estimate of the measurand is incomplete without a statement of uncertainty surrounding the estimate. Figure 1.2 illustrates the concepts of MPE and uncertainty as they pertain to measurements on features with various lengths. In this plot, the error of each measurement result from the corresponding reference value is plotted as a function of the measured length. Some of the measurement points consist of uncertainty intervals that exceed the MPE (dotted line). The use of these measurement results in deciding whether or not the instrument satisfies the MPE would prove inconclusive. Finally, unlike the MPE of a measuring instrument, measurement uncertainty is a task-specific quantity.

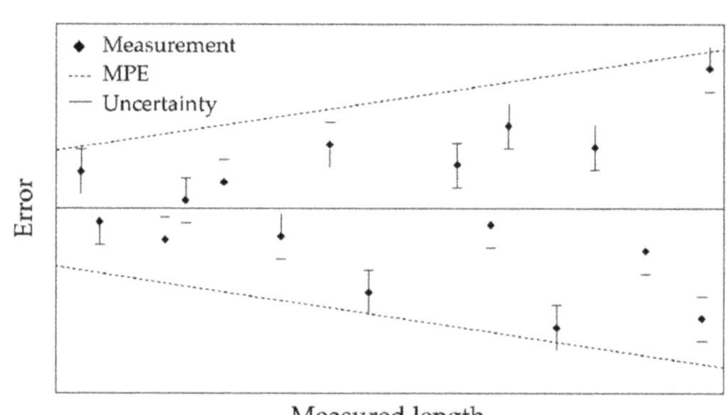

Figure 1.2. Plot of MPE for a CMS, also showing how uncertainty can be expressed.

1.4 Specification standards for coordinate metrology

1.4.1 Coordinate metrology standards

The ISO 10360 series of specification standards defines the terminology, instrument types and procedures for performance verification of CMSs. The series is broken down into seven parts, which are listed below and described in detail in Hocken and Pereira (2016). The ISO 10360 framework was originally developed for contact coordinate measuring machines (CMMs), but has been (or is being) extended in recent years to optical and X-ray computed tomography based CMSs.

- Part 1: Vocabulary (ISO 2000a).
- Part 2: CMMs used for measuring linear dimensions (ISO 2009).
- Part 3: CMMs with the axis of a rotary table as the fourth axis (ISO 2000b).
- Part 4: CMMs used in scanning measuring mode (ISO 2000c).
- Part 5: CMMs using single and multiple-stylus contacting probing systems (ISO 2010b).
- Part 6: Estimation of errors in computing Gaussian associated features (ISO 2001).
- Part 7: CMMs equipped with imaging probing systems (ISO 2011).
- Part 8: CMMs with optical distance sensors (ISO 2013a).
- Part 9: CMMs with multiple probing systems (ISO 2013b).
- Part 10: Laser trackers for measuring point-to-point distances (ISO 2016a).
- Part 11: CMMs using the principle of computed tomography (ISO/CD 2020a)—still at committee draft stage.
- Part 12: Articulated arm coordinate measurement machines (ISO 2016b).
- Part 13: Optical 3D CMS (ISO 2020b)—still at committee draft stage.

The parts of ISO 10360 most relevant to optical CMSs are parts 8, 10 and 13. Part 8 covers CMSs that use distance measuring sensors, defined as non-contacting probing systems which determine a corrected measured point by means of an optical distance measurement principle. Examples given in part 8 are triangulation and coaxial distance measurement. The former includes fringe projection (chapter 5), Moiré interferometry, slit light projection and point scanning, and the latter includes interferometry and confocal systems. The basic principles of these methods are covered in Harding (2014), Zhang (2013) and Leach (2011) and advances are the focus of this book. Part 13 is still at the draft stage at the time of writing, but will cover optical 3D CMS—the distinction with part 8 is that the systems in part 13 are stand-alone CMSs, they are not CMS platforms with optical sensors. This distinction is, however, somewhat ambiguous, for example, which part should be applied when using a rotation stage to position the object being measured with a fringe projection system? Chapter 8 will discuss this distinction and ambiguity. Part 10 covers CMSs in which a cooperative target is followed with a laser beam and its location determined in terms of range and two angles, i.e. CMSs using laser trackers. Advances in laser trackers are not covered in this book, but see Harding (2014) and Schmitt *et al* (2016).

Acknowledgements

Thanks to Dr Adam Thompson (University of Nottingham) for helping me with the intricacies of MS Word.

References

BIPM, IEC, IFCC, ILAC, ISO, IUPAC, IUPAP and OIML 2012 *International Vocabulary of Metrology—Basic and General Concepts and Associated Terms—JCGM 200* (Sèvres: Bureau International des Poids et Mesures)

Haitjema H 2018 Measurement uncertainty *Basics of Precision Engineering* ed R K Leach and S T Smith (Boca Raton, FL: CRC Press)

Harding K 2014 *Handbook of Optical Dimensional Metrology* (Boca Raton, FL: CRC Press)

Hocken R J and Pereira P H 2016 *Coordinate Measuring Machines and Systems* (Boca Raton, FL: CRC Press)

ISO 10360 part 1 2000a *Geometrical Product Specifications (GPS)—Acceptance and Reverification Tests for Coordinate Measuring Machines (CMM)—Part 1: Vocabulary* (Geneva: International Organization for Standardization)

ISO 10360 part 3 2000b *Geometrical Product Specifications (GPS)—Acceptance and Reverification Tests for Coordinate Measuring Machines (CMM)—Part 3: CMMs with the axis of a rotary table as the fourth axis* (Geneva: International Organization for Standardization)

ISO 10360 part 4 2000c *Geometrical Product Specifications (GPS)—Acceptance and Reverification Tests for Coordinate Measuring Machines (CMM)—Part 4: CMMs used in scanning measuring mode* (Geneva: International Organization for Standardization)

ISO 10360 part 6 2001 *Geometrical Product Specifications (GPS)—Acceptance and Reverification Tests for Coordinate Measuring Machines (CMM)—Part 6: Estimation of errors in computing Gaussian associated features* (Geneva: International Organization for Standardization)

ISO 17025 2005 *General Requirements for the Competence of Testing and Calibration Laboratories* (Geneva: International Organization of Standardization)

ISO 10360 part 2 2009 *Geometrical Product Specifications (GPS)—Acceptance and Reverification Tests for Coordinate Measuring Machines (CMM)—Part 2: CMMs used for measuring size* (Geneva: International Organization for Standardization)

ISO 10110 part 8 2010a *Optics and Photonics—Preparation of Drawings for Optical Elements and Systems—Part 8: Surface texture; roughness and waviness* (Geneva: International Organization for Standardization)

ISO 10360 part 5 2010b *Geometrical Product Specifications (GPS)—Acceptance and Reverification Tests for Coordinate Measuring Machines (CMM)—Part 5: CMMs using single and multiple-stylus contacting probing systems* (Geneva: International Organization for Standardization)

ISO 10360 part 7 2011 *Geometrical Product Specifications (GPS)—Acceptance and Reverification Tests for Coordinate Measuring Machines (CMM)—Part 7: CMMs equipped with imaging probing systems* (Geneva: International Organization for Standardization)

ISO 10360 part 8 2013a *Geometrical Product Specifications (GPS)—Acceptance and Reverification Tests for Coordinate Measuring Machines (CMM)—Part 8: CMMs with optical distance sensors* (Geneva: International Organization for Standardization)

ISO 10360 part 9 2013b *Geometrical Product Specifications (GPS)—Acceptance and Reverification Tests for Coordinate Measuring Machines (CMM)—Part 9: CMMs with multiple probing systems* (Geneva: International Organization for Standardization)

ISO 10360 part 10 2016a *Geometrical Product Specifications (GPS)—Acceptance and Reverification Tests for Coordinate Measuring Machines (CMM)—Part 10: Laser trackers for measuring point-to-point distances* (Geneva: International Organization for Standardization)

ISO 10360 part 12 2016b *Geometrical Product Specifications (GPS)—Acceptance and Reverification Tests for Coordinate Measuring Machines (CMM)—Part 12: Articulated arm coordinate measurement machines (CMM)* (Geneva: International Organization for Standardization)

ISO/CD 10360 part 11 2020a *Geometrical Product Specifications (GPS)—Acceptance and Reverification Tests for Coordinate Measuring Machines (CMM)—Part 11: CMMs using the principle of computed tomography (CT)* (Geneva: International Organization for Standardization)

ISO/CD 10360 part 13 2020b *Geometrical Product Specifications (GPS)—Acceptance and Reverification Tests for Coordinate Measuring Machines (CMM)—Part 13: Optical 3D CMS* (Geneva: International Organization for Standardization)

Leach R K 2011 *Optical Measurement of Surface Topography* (Berlin: Springer)

Leach R K 2014 *Fundamental Principles of Engineering Nanometrology* 2nd edn (Amsterdam: Elsevier)

Leach R K 2015 Surface texture *CIRP Encyclopedia of Production Engineering* ed L Laperrière and G Reinhart (Berlin: Springer)

Leach R K, Bourell D, Carmignato S, Donmez A, Senin N and Dewulf W 2019 Geometrical metrology for metal additive manufacturing *Ann. CIRP* **68** 677–700

Leach R K, Giusca C L, Haitjema H, Evans C and Jiang X 2015 Calibration and verification of areal surface texture measuring instruments *Ann. CIRP* **64** 797–813

Leach R K and Shaheen A 2020 Post-process coordinate metrology ed R K Leach and S Carmignato *Precision Metal Additive Manufacturing* (Boca Raton, FL: CRC Press)

Schmitt R H, Peterek M, Morse E, Knapp W, Galetto M, Härtig F, Goch G, Hughes B, Forbes A and Estler W T 2016 Advances in large-scale metrology—review and future trends *Ann. CIRP* **65** 643–65

Stavroulakis P and Leach R K 2016 Review of post-process optical form metrology for industrial-grade metal additive manufactured components *Rev. Sci. Instrum.* **87** 041101

Zhang S 2013 *Handbook of 3D Machine Vision: Optical Metrology and Imaging* (Boca Raton, FL: CRC Press)

Chapter 2

State-of-the-art in point cloud analysis

Sofia Catalucci and Nicola Senin

Three-dimensional point clouds play a central role in industrial metrology—they are the product of coordinate measurement systems and the starting point for the data processing methods that produce inspection and verification results. As measurement technologies evolve, increasingly dense point clouds are being produced, capable of capturing geometric information at unprecedented levels of detail, which is particularly important for the complex freeform and multi-scale surfaces typically found in additively manufactured parts. In this chapter, the state-of-the-art in point cloud analysis and processing is presented, covering formal representation, pre-processing, partitioning, fitting and registration. Recent literature on the incorporation of measurement error and measurement uncertainty into point cloud analysis and processing is also discussed.

2.1 Introduction

In this chapter, the term 'point cloud' is used to refer to a point cloud of surface data, i.e. a set of points representative of the surface of a part. A point cloud is, therefore, seen as a primitive representation of a three-dimensional (3D) object (figure 2.1), identified as a set of points in a coordinate system (Rusu and Cousins 2011, Aldoma *et al* 2012, Saval-Calvo *et al* 2015).

In a Cartesian coordinate system, a point cloud is defined by the x, y and z coordinates of each specific point. Due to their simplicity, point clouds have been introduced into a variety of areas related to geometric reconstruction, processing, modelling and rendering. Point clouds are considered unstructured sets, as the points are stored with no information concerning topological connectivity (Alexa *et al* 2004, Kobbelt and Botsch 2004). Point clouds can only provide an approximation of 3D geometry, because of the intrinsic nature of point-based samplers (Alexa *et al* 2004), a problem exacerbated when points are taken on curved surfaces.

doi:10.1088/978-0-7503-2524-0ch2

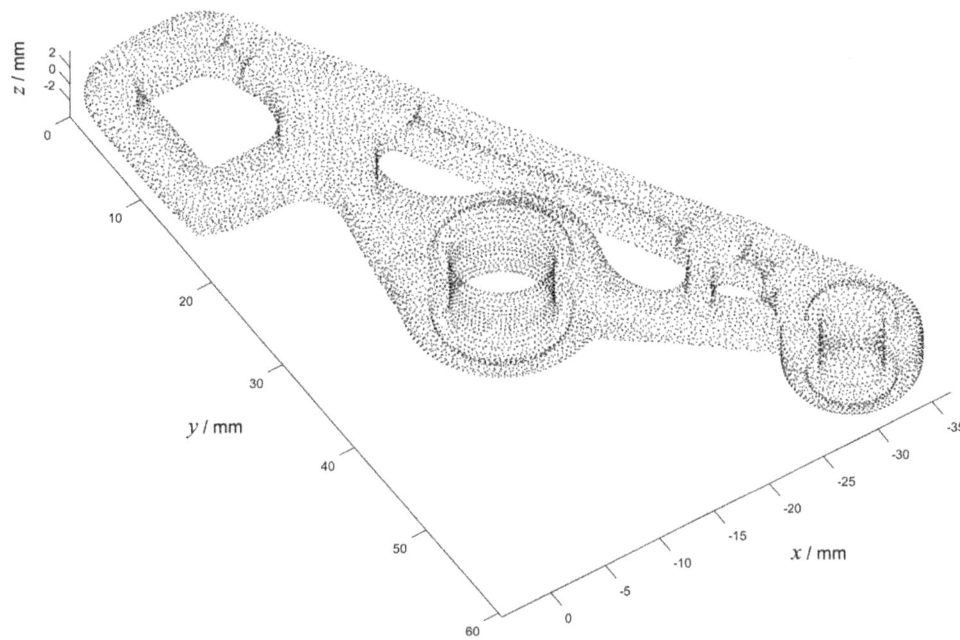

Figure 2.1. Point cloud obtained by optical measurement of an automotive part (roof bracket; part designed by the Centre for Additive Manufacturing, University of Nottingham).

2.1.1 Mathematical representation of 3D point clouds

A point cloud in Cartesian space is generally represented as an $n \times 3$ matrix:

$$\begin{bmatrix} x_1 & y_1 & z_1 \\ x_2 & y_2 & z_2 \\ \vdots & \vdots & \vdots \\ x_n & y_n & z_n \end{bmatrix}, \tag{2.1}$$

where n is the number of points and each point is represented by its three coordinates x, y and z (Sun *et al* 2016). Within a point cloud, additional point-related attributes may also be stored (Weinmann 2016), for example curvature, which is then visually mapped to an RGB colour space (see figure 2.2). When used to represent surfaces, point clouds may also store the local normal vector of the surface in correspondence to each sampled point, i.e. an additional set of three scalar values associated with each row of the matrix defined in equation (2.1). The mathematical representation of a point cloud can, therefore, be generalised into a collection of multidimensional points (Rusu and Cousins 2011), where the number of dimensions is the number of spatial coordinates plus any additional attribute which needs to be recorded for each point.

2.1.2 Common 3D point cloud file formats

A point cloud can be stored in a file using several formats (Landa *et al* 2013). All formats must accommodate the x, y and z scalar values associated with each point,

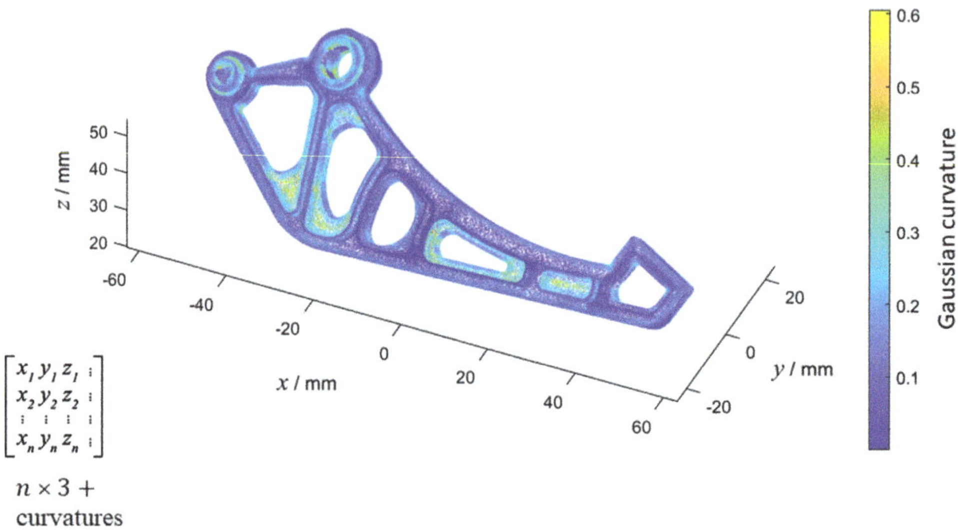

Figure 2.2. Point cloud augmented with Gaussian curvature information. Curvature is visualised by mapping curvature values to a RGB colour map.

as well as any additional information that may be stored for the points. File formats for point clouds are further classified based on how structured information is encoded in the file, i.e. text or binary formats. In text files, numerical values are first converted into sequences of characters, then characters are encoded into sequences of bits using either ASCII or UNICODE schemes. In binary files, numerical values are directly converted into sequences of bits using dedicated encoding schemes. The advantage of text files is that they can be opened by any editor and directly interpreted by humans. In contrast, binary files require knowledge of the encoding/decoding scheme originally used to create them, in order to reconvert sequences of bits back into numbers. The main disadvantages of text files are:

(a) *Size*: a dense point cloud may occupy a large amount of disk space, and be inefficient to read and write.
(b) *Accuracy*: a choice must be made on how to convert a real number (for example, a coordinate) into a sequence of characters of finite length; using too few characters may lead to truncation errors.

Binary files are more compact and efficient to write and read, although dedicated software implementing the selected encoding/decoding scheme must be available, which reduces portability across different software platforms. Common point cloud file formats are listed in table 2.1.

2.1.3 Point cloud transformations

Point cloud transformations fall into two categories: rigid transformations and non-rigid transformations (figure 2.3).

Table 2.1. Common point cloud file formats.

Filename extension	Information encoding	File structure	Contents	Notes and references
XYZ	Text	Points list (lists of other elements, i.e. RGB)	Non-standardised set of files based on x, y and z Cartesian coordinates.	
ASC	Text	Points list (lists of other elements, i.e. RGB)	Non-standardised set of files based on x, y and z Cartesian coordinates.	
PLY	Text and binary	Header Points list Face list (lists of other elements, i.e. RGB, normals)	x, y and z Cartesian coordinates and a series of user-selected properties.	Designed either to handle triangle meshes and to store point clouds (Polygon file format or Stanford Triangle format (Turk 1994)).
PCD	Binary	Version Fields Size Type Count Width Height Viewpoints No. points Points list	Fields: x, y and z Cartesian coordinates and a series of user-selected properties.	Point Cloud Data file format, used inside Point Cloud Library PCL (Rusu and Cousins 2011).
LAS	Binary	No. points Points list (lists of other elements, i.e. RGB, NIR, GPS time)	x, y and z Cartesian coordinates and a series of user-selected properties.	LiDAR Aerial Survey (LASer) format, specified by the American Society for Photogrammetry and Remote Sensing.

PTX	Text	Number of rows Number of columns Scanner registered position Scanner registered axes Transformation matrix Points list (lists of other elements, i.e. RGB)	The header contains the scanner location and any transformation matrices to transform multiple scans into one coordinate system; this is followed by x, y and z data, and RGB colour information.	Typically, from LIDAR scanners and Leica Cyclone (Plain Text).
PTS	Text	No. points Points list (lists of other elements, i.e. intensity values, RGB information)	The first line gives the number of points to follow. Each subsequent line has seven values, the first three are the x, y and z Cartesian coordinates, the fourth is an 'intensity' value, and the last three are the RGB colour information.	Typically, from LIDAR scanners and Leica Cyclone (laser scan plain data format).
E57	Binary	Header Binary sections (encoding the point data or images) XML section	x, y and z Cartesian coordinates and a series of user-selected properties, such as normals, colours and scalar field intensity.	Suitable for storing images and metadata produced by laser scanners and other 3D imaging systems (ASTM E57 Committee on 3D Imaging Systems (Huber 2011)).
OBJ	Text and binary	Geometric vertices (v) Texture vertices (vt) Vertex normals (vn) Face list (f)	x, y and z Cartesian coordinates and a series of user-selected properties, such as texture, normals, faces; OBJ coordinates have no units, but OBJ files can contain scale information.	Wavefront technologies; suitable for 3D geometries; an OBJ file may contain vertex data, elements, freeform curve/surface body statements, connectivity between freeform surfaces, grouping and display/render attribute information.
Others: FLS FBX			(FARO Cloud) binary text and binary	

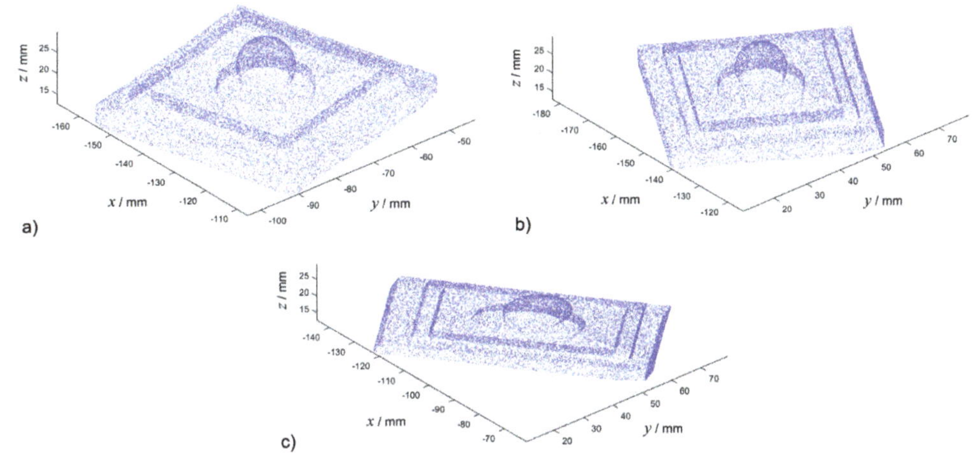

Figure 2.3. Example of rigid and non-rigid transformations applied to a point cloud representing a test artefact (Manufacturing Metrology Team, University of Nottingham): (a) original cloud, (b) transformed by rotation of 45 degrees along the z-axis (rigid transformation), (c) stretched along the x-axis by selective rescaling of the x-coordinates (non-rigid transformation).

Rigid transformations result in translations and/or rotations of the point cloud with respect to the local coordinate system, or equivalently, in changes of coordinate system. Non-rigid transformations instead imply a change in shape/size of the cloud, i.e. a change in relative positions between points (Weckenmann *et al* 2009, Tam *et al* 2013). Examples of non-rigid transformations include bending or stretching of a point cloud or changing its scale. Rigid and non-rigid transformations are typically represented using a 4 × 4 matrix, meant to be pre-multiplied or post-multiplied (depending on the convention) with each point in the point cloud, where points are expressed in homogeneous coordinates. The transformation, in the pre-multiplication format, can be written in vector form,

$$\boldsymbol{q}_i = \mathbf{M}\boldsymbol{p}_i, \tag{2.2}$$

and in scalar form,

$$\begin{bmatrix} x^q{}_i & y^q{}_i & z^q{}_i & 1 \end{bmatrix} = \begin{bmatrix} m_{11} & m_{12} & m_{13} & m_{14} \\ m_{21} & m_{22} & m_{23} & m_{24} \\ m_{31} & m_{32} & m_{33} & m_{34} \\ m_{41} & m_{42} & m_{43} & m_{44} \end{bmatrix} \cdot \begin{bmatrix} x^p{}_i & y^p{}_i & z^p{}_i & 1 \end{bmatrix}^T, \tag{2.3}$$

where $\boldsymbol{q}_i = \begin{bmatrix} x^q{}_i & y^q{}_i & z^q{}_i \end{bmatrix}$ is the ith point after the transformation, $\boldsymbol{p}_i = \begin{bmatrix} x^p{}_i \, y^p{}_i \, z^p{}_i \end{bmatrix}$ is the original ith point (i.e. before the transformation) and $\mathbf{M}$ is the 4 × 4 transformation matrix. In the case that the transform is rigid (rotation and translation), the 4 × 4 transformation matrix $\mathbf{M}$ is typically expressed as

$$\begin{bmatrix} r_{11} & r_{12} & r_{13} & t_x \\ r_{21} & r_{22} & r_{23} & t_y \\ r_{31} & r_{32} & r_{33} & t_z \\ 0 & 0 & 0 & 1 \end{bmatrix}, \tag{2.4}$$

where r_{ij} are the components of the 3×3 rotation matrix and t_x, t_y, t_z are the components of a 3×1 translation vector. In the case of rigid transformations, another common notation is the following

$$\boldsymbol{q}_i = \mathbf{R}\boldsymbol{p}_i + \boldsymbol{t}, \tag{2.5}$$

where the rotation matrix and translation vector embedded in the transformation matrix in equation (2.4) are separated so that the $\mathbf{R}$ is a 3×3 rotation matrix, $\boldsymbol{t}$ is a 1×3 translation vector and point $\boldsymbol{p}_i$ (before) and $\boldsymbol{q}_i$ (after the transformation) do not need to be expressed in homogeneous coordinates.

In industrial measurement applications, point coordinates are the result of a measurement process, thus any transformation must be considered with care in relation to how it may alter the measurement results. Rigid transformations are primarily required when the point cloud needs to be referred to a different coordinate system, or equivalently, when the point cloud needs to be aligned to another geometric object, typically another point cloud (see section 2.6 for point cloud registration). Non-rigid transformations may instead be applied to correct scaling error, or to compensate for measurement-induced distortion.

2.2 Extracting properties and organising information in point clouds

Several methods exist to analyse point clouds and extract their salient properties. The challenges of navigating a topologically unstructured set, such as a point cloud, can be eased by introducing organisation, for example through spatial grouping by proximity, or by computing geometric properties mapped to subsets (regions) of the cloud. Efforts to add organisation and structure to point clouds lead to easier navigation, easier and quicker searching, and retrieval of points with specific properties, easier and quicker processing (filtering, aligning) and more effective comparison when multiple point clouds are involved.

2.2.1 Convex hull and bounding boxes

The computation of convex hull and bounding boxes is used to obtain a quick estimate of spatial occupancy and boundaries of a point cloud. A set of points in a Euclidean space (i.e. a point cloud) is defined to be convex if it contains the line segments connecting each pair of its points. The convex hull of a point cloud $\mathbf{P}$ (convex or not) is the smallest convex set containing all the points, mathematically the intersection of all convex sets containing the points.

In computational geometry, given a set $\mathbf{P}$ of n points $\boldsymbol{p}_i$, $\forall\, i \in \{1, \ldots, n\}$, the linear combination of the points is given by equation (2.6) and is called a convex combination (De Berg *et al* 2008)

$$\sum_{i=1}^{n} \lambda_i \boldsymbol{p}_i, \tag{2.6}$$

where λ_i satisfies the conditions $\lambda_i \geqslant 0\ \forall\ i$ and $\sum_{i=1}^{n} \lambda_i = 1$. The convex hull of a set of points is exactly the set of all possible convex combinations of the points.

The convex hull of a finite point set $\mathbf{P} \subset \mathbf{R}^d$ forms a convex polygon when $d = 2$ and a convex polyhedron when $d = 3$. A convex polygon (or polyhedron) is specified by the sets of its vertices and the edges connecting them. In 3D space, the convex hull is the boundary of a closed convex surface generated by applying Delaunay triangulations with the outer points, vertices of the smallest possible convex bounding volume identified by a triangle mesh (figure 2.4).

A bounding box is a parallelepiped in 3D, or a rectangle in 2D space, enclosing a geometric object. When the box is aligned with the axes of the coordinate system, it is then known as an axis-aligned bounding box (AABB—see example in figure 2.5).

To distinguish an arbitrarily oriented bounding box from an AABB, the former is called an oriented bounding box (OBB) (Schneider and Eberly 2003). The minimum or smallest bounding box for a point set **P** is the box with the smallest measure (area, volume or hypervolume in higher dimensions) within which all the points of **P** lie. The minimum bounding box of a point set is the same as the minimum bounding box of its convex hull.

2.2.2 Centroid and principal axes of a point cloud

The centroid $\boldsymbol{c}$ of a point cloud **P** (in 2D or 3D) is a point whose coordinates are defined as the arithmetic mean of the coordinates of all the points of the cloud (Abdi 2009). The centroid is often used as a first, coarse indicator of point cloud localisation in space (central tendency).

In addition to knowing a point cloud localisation, it is important to determine how points are dispersed (i.e. scattered) in space about the centroid. Dispersion can be measured along any arbitrary axis passing through the centroid. However, it is generally convenient to evaluate dispersion along specific directions known as principal axes (figure 2.6).

The principal axes are defined as orthogonal directions (two in 2D space or three in 3D space) where the largest, intermediate and smallest dispersion of the points can be observed respectively (James *et al* 2013, Tam *et al* 2013). Principal axes are typically found by running principal component analysis (PCA) (Jolliffe 2002).

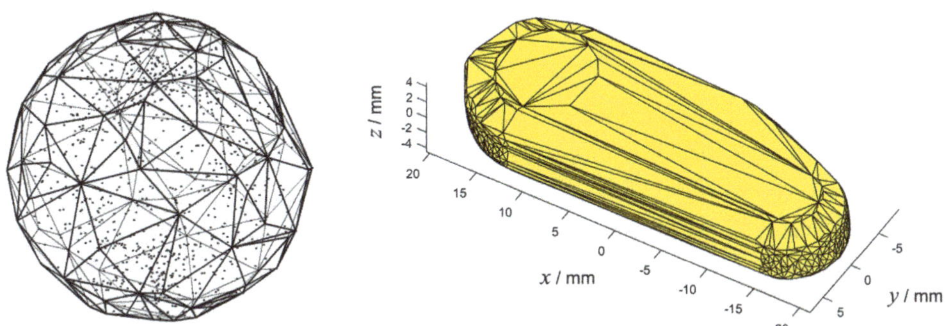

a) b)

Figure 2.4. Convex hulls obtained by Delaunay triangulation: (a) for a random set of points and (b) for the roof bracket part.

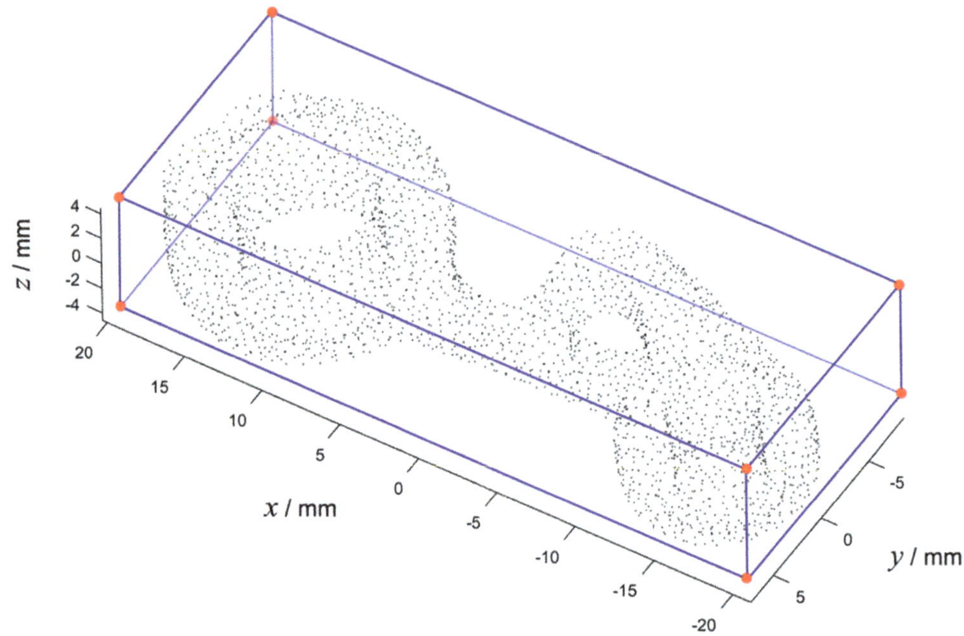

Figure 2.5. Axis-aligned bounding box, enclosing a point cloud (simplified cloud from measurement of the roof bracket).

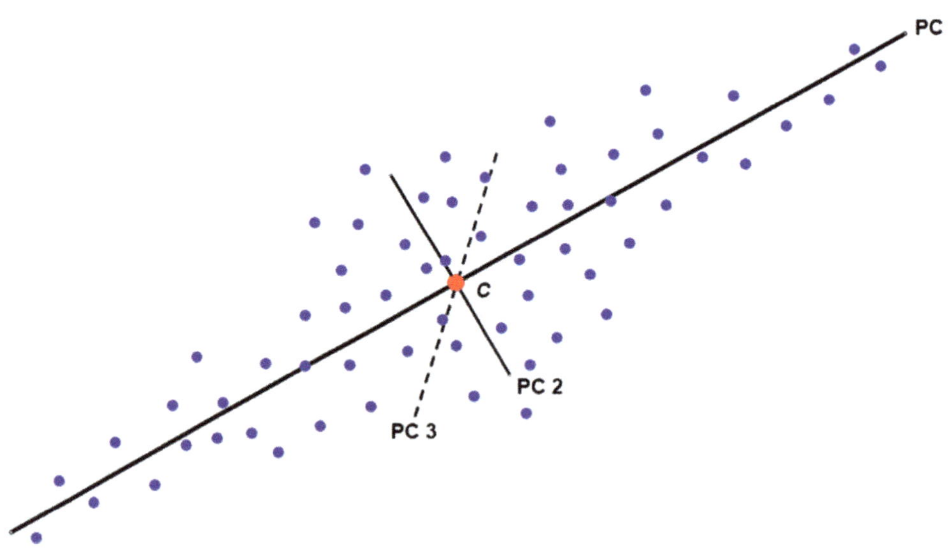

Figure 2.6. Example result of principal component analysis for a point cloud; centroid *c* (red) and first (PC1), second (PC2) and third (PC3) principal axes (black) for an example point cloud (blue).

In statistical analysis, PCA is used to simplify random variables of high dimensionality by identifying a smaller set of new dimensions (linear combinations of the original ones) capable of encapsulating most of the variance for the original set. In geometric modelling, PCA can be computed on a point cloud by considering each point as a realisation of a tri-variate random variable (bivariate if in 2D space). The principal components, found from the eigenvectors and eigenvalues of the covariance matrix, correspond to axes in space (2D or 3D) passing through the centroid, and are called principal axes.

Given a set of points $\mathbf{P} = \{p_1, ..., p_n\}$ with centroid $\boldsymbol{c}$, the covariance matrix $\mathbf{C}$ is built as follows:

$$\mathbf{C} = \mathbf{P} \times \mathbf{P}^{\mathrm{T}}, \tag{2.7}$$

where $\mathbf{P}$ is expressed as

$$\mathbf{P} = \begin{bmatrix} p_{1x} - c_x & p_{2x} - c_x & \cdots & p_{nx} - c_x \\ p_{1y} - c_y & p_{2y} - c_y & \cdots & p_{ny} - c_y \\ p_{1z} - c_z & p_{2z} - c_z & \cdots & p_{nz} - c_z \end{bmatrix}, \tag{2.8}$$

whose ith column is vector $\boldsymbol{p}_i - \boldsymbol{c}$.

Centroid and principal axes are often used as coarse descriptors of localisation and orientation of a point cloud and are, therefore, often used as the starting point to solve alignment problems involving multiple point clouds (see section 2.6.1.2). In addition, if the point cloud is representative of a portion of surface with uniform spatial orientation (for example, a plane or a surface with small curvature), principal axes can be used to identify the orientation of the local surface normal, corresponding to the principal axis capturing the least variance (see section 2.4.1). Since PCA also allows computation of the amount of dispersion along the principal axes through the eigenvalues of the covariance matrix, information about the cloud aspect ratio can also be obtained (for example, rate of elongation with respect to one or two principal axes) paving the way for methods for shape classification and comparison.

2.2.3 Spatial subdivision of point clouds

A way to add organisation to a topologically unstructured set of points is to subdivide the set into groups of points that occupy similar positions in space (Schneider and Eberly 2003, De Berg *et al* 2008). Spatial subdivision is an efficient way for enabling fast search and retrieval of points (for example, to find the closest points to a given geometric reference—point, plane, another point cloud, etc). Many spatial subdivision methods are hierarchical, meaning that subdivision is performed recursively, leading to the generation of nested volumes (or areas in 2D) of increasingly smaller occupancy, each containing points from the original point cloud. Hierarchical subdivisions give origin to graph-like structures (trees), where nodes contain the description of each subset generated by the subdivision, including the points they contain. Hierarchical subdivision trees can be navigated across levels (an operation known as 'tree traversal'), typically performing recursive searchers starting from coarse subdivisions and moving towards finer ones, in order to retrieve points. Most of the hierarchical, recursive subdivision methods discussed in this section can be adapted to operate on other geometric primitives in addition to points, for example triangle meshes (see section 2.3.3). Also, most of the analytical definitions for recursive spatial subdivisions work in *d*-dimensional spaces, meaning that they have identical formulations in 2D, 3D or any space of higher dimensionality.

2.2.3.1 AABB trees, binary space partitioning and kd-trees

AABB trees (van den Bergen 1997) adopt a hierarchical, recursive decomposition where each generated convex subset of space is an axis-aligned bounding box (AABB, as discussed in section 2.2.1). A multitude of different AABB trees can be implemented depending on the method adopted to partition space, how many partitions are generated per decomposition level, whether partitions are overlapping or not and whether partitions contain only points. The freedom in the decomposition procedure makes AABB trees particularly convenient in applications involving collision detection of complex geometries, because custom trees can be built to address specific requirements of specific regions in the geometry.

Binary space partitioning (BSP) encompasses a family of recursive decomposition techniques where the common denominator is that each subdivision is strictly binary, meaning that it divides the *d*-dimensional space into two partitions only, known as half-spaces (Jiménez *et al* 2001), and the partitions are non-intersecting. The subdivision is typically performed by a hyperplane, whose position and orientation depends on type of BSP. At each level, the partitioning is binary, therefore, in the resulting tree, each node (corresponding to a partition) generates only two children ('leaves'—the result of the binary subdivision of the partition).

The most popular type of BSP is the kd-tree (Bentley 1975, Samet 1989, 1990). An example of kd-tree computed on a triangle mesh is shown in figure 2.7.

The kd-tree is the result of a very specific type of recursive spatial partitioning where:

(a) the partitioning at each decomposition level is binary (i.e. the decomposition is a type of BSP);
(b) the generated partitions are AABBs (i.e. the generated tree is a AABB tree); and
(c) the position of the partitioning hyperplane is decided by very specific criteria, described in the following.

In kd-tree decomposition for each convex AABB set to be partitioned, a hyperplane is created aligned to one of the three main axes (x, y or z), usually the one corresponding to maximum point dispersion within the partition. Then, the position of the hyperplane is set as coincident to the median value of the coordinates of the points along the same axis. This default partitioning behaviour generates balanced kd-trees (balanced—each point is reachable by traversing approximately the same number of nodes starting from the root). However, different hyperplane positioning strategies can be implemented depending on the application.

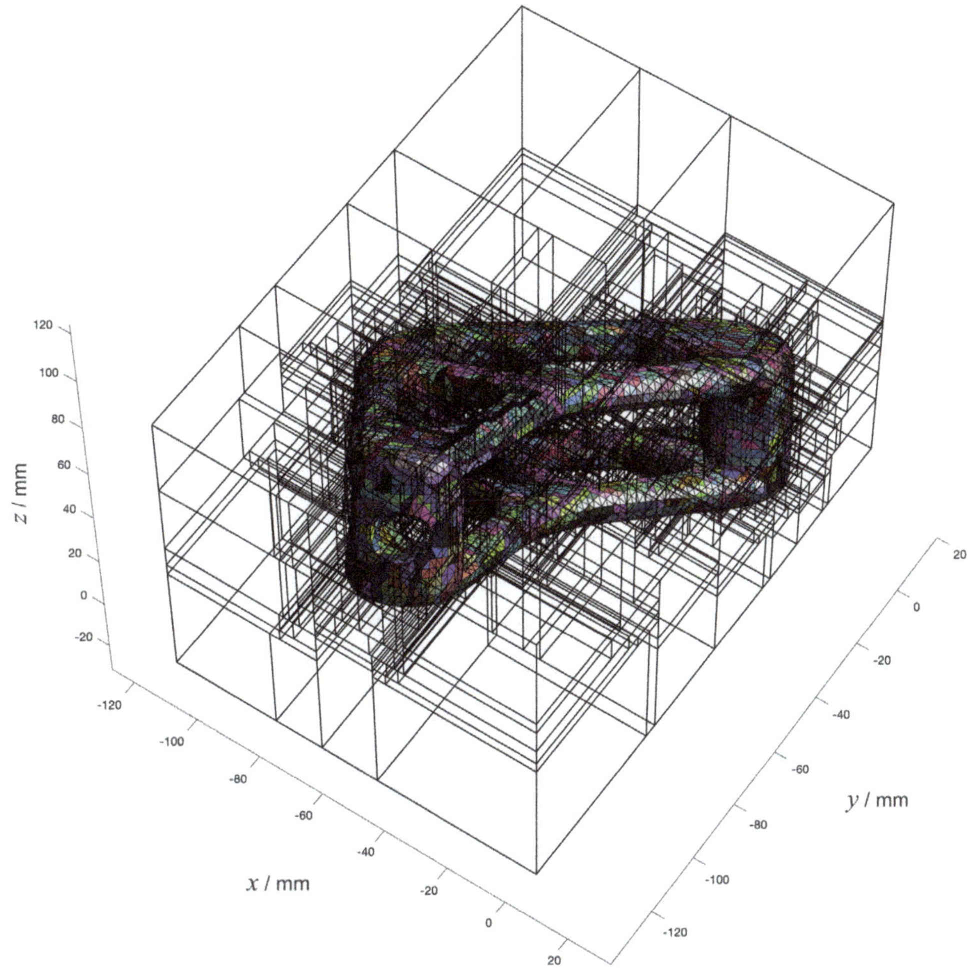

Figure 2.7. Example of kd-tree decomposition applied to a triangle mesh; the mesh has been artificially generated and is representative of typical hollow structures which could be produced by additive manufacturing.

2.2.3.2 Octree decomposition

Another popular hierarchical subdivision method is known as octree decomposition (Schneider and Eberly 2003, De Berg *et al* 2008, Aluru 2004). In octree decomposition, each spatial partition is described as a 3D cuboid (again a type of AABB). Recursive partitioning consists of subdividing the cuboid into eight equal parts (also cuboids) known as octants. Note that cuboids degenerate into cubes if the root AABB has the same sizes along the three main axes. Analogously to kd-trees, octree decompositions use hyperplanes and create convex, non-intersecting partitions. However, eight new partitions are generated from each decomposed AABB (creating a tree with eight leaves per node) and the position of the hyperplanes cannot be optimised due to the requirement for generating eight equal cuboids. Due to the generation of octants, octree decompositions are applicable to 3D geometries only. The 2D equivalent of an octree decomposition is known as quadtree decomposition. In this case, each partition is a 2D AABB shaped as a rectangle (or cube if the sizes are equal), and each decomposition generates four equal rectangles, known as quadrants (i.e. four leaves per node). The same considerations formulated when comparing octree decompositions to kd-trees apply to quadtree decompositions, compared to 2D kd-trees.

2.2.3.3 Cell decompositions

In cell decompositions, the point cloud (or other geometrical object) is subdivided into a set of 3D cells (in 3D space) or 2D cells (in 2D space) of equal volume/area. There is no recursive subdivision and cells are immediately generated at their final size (Shapiro 2002). Each cell is filled with the points located within the occupied region. In the most common case, cells are cubic. In such cases, they are referred to as voxels, and cell decomposition is called voxelisation. The use of cubic cells brings several advantages, the main one being that cells can be organised in a uniform 2D or 3D grid. In the grid, cells are indexed and can be easily navigated, for example paving the way to the use of data manipulation technologies derived from image processing, though in three dimensions. Voxelisations are very simple to obtain. The most common approach is to create a uniform grid of the desired size on top of the cloud, and then iterate through the cells in order to identify the points falling within each cell (Hosoi and Omasa 2006). Voxelisations have the main advantage of creating a topological infrastructure associated with the point cloud, because adjacency between points can be inferred directly from voxel indices in the grid. However, the decomposition is often inefficient—voxels that are too large may contain many points, resulting in an excessively coarse decompositions; in contrast, very small cells may contain no points in low density regions of the point cloud, leading to a waste of memory and computational challenges in search/retrieval. Abandoning constant voxel size in favour of variable-sized cells means losing the advantages of dealing with uniform grids and thus is seldom implemented.

2.2.3.4 Point search and retrieval using spatial subdivisions of point clouds

The main reason to perform spatial subdivision of a point cloud is related to the generation of a structured organisation in an otherwise topologically unstructured dataset. The organisation into a structure facilitates and accelerates search and retrieval operations, which ultimately consist of finding the nearest points to a given set of coordinates, either in the same point cloud, or in an external entity (Rusu 2009). In other words, given a set of points $\mathbf{P}$ and a query point $\boldsymbol{q}$, which may or may not belong to $\mathbf{P}$, the problem is to find a subset of $\mathbf{P}$ containing either the k closest points to $\boldsymbol{q}$ (figure 2.8(a)), or all the points whose Euclidean distance to $\boldsymbol{q}$ is smaller than a given radial distance r (figure 2.8(b)).

In the absence of a spatial subdivision, the only way to solve the aforementioned tasks is by brute force, by computing the Euclidean distances of all the points in $\mathbf{P}$ to the query point $\boldsymbol{q}$. This is done, for example, by the most popular search and retrieval algorithm for point clouds, known as the k nearest neighbours algorithm (k-NN). In k-NN, once the Euclidean distances to three query points are known, points can be ranked on increasing distance so that the k nearest ones can be returned (Rusu 2009, Elseberg *et al* 2012). A similar approach is adopted to identify points at a maximum distance r from the query point $\boldsymbol{q}$. Again, once all the Euclidean distances to $\boldsymbol{q}$ are computed, the selection is driven by thresholding on the given radial distance r.

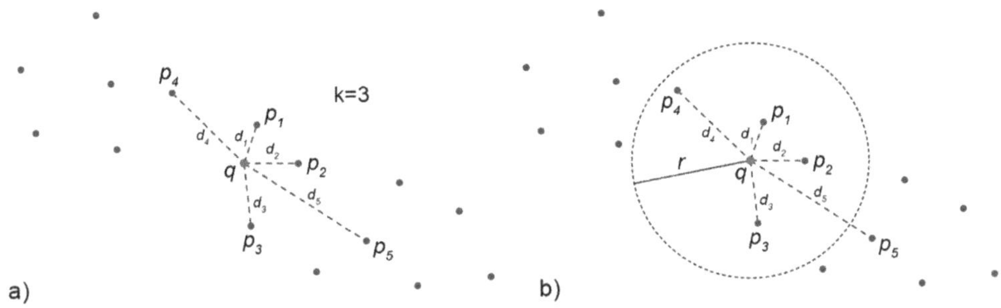

Figure 2.8. Schematic representation of the two most typical point search and retrieval problems: (a) find the k nearest neighbours of a given point $\boldsymbol{q}$; (b) find all the points that lay within a radius r from $\boldsymbol{q}$.

In contrast, if a spatial subdivision is available, the same procedures to retrieve the k nearest points or all the points within a given radial distance r can be applied on much smaller sets of points, identified from selected nodes in the subdivision tree (the node selection procedure depends on the type of subdivision, but essentially consists of traversing the tree until some conditions are met).

Spatial subdivision can also be used to navigate across points of a point set using the Euclidean minimum spanning tree (EMST) method (Hoppe *et al* 1992). In the EMST method, a path is automatically found to connect any two arbitrarily chosen points of a set. The path is formed by finding a sequence of points to be connected by segments, so that the total path length is minimised. Each segment represents the Euclidean distance between the pair of points it connects. Spatial subdivision methods greatly help to speed up the identification of which points to connect in the EMST.

2.3 Point cloud pre-processing

Many industrial coordinate measurement technologies produce high-density point clouds. In the context of inspection and verification, high-density implies high-resolution spatial sampling, and thus potentially richer information available to assess whether the part conforms to dimensional and geometric specification requirements, in particular for smaller-sized features. Nevertheless, several issues typically affect the quality of high-density point clouds obtained by measurement—part coverage may be suboptimal, with some surfaces being suitably represented by truly high-density sampling, while other surfaces may be less covered, or not represented at all (for example, if the surface is not within line-of-sight of an optical instrument, or unreachable by a contact probe). On the other hand, some surfaces may be subjected to excess/redundant sampling. Excess sampling may occur, for example, in correspondence to regions that have been sampled multiple times as a result of partially overlapping fields of view. Other problems affecting high-density point clouds, in addition to sampling coverage, are related to measurement error leading to point misplacement. Typical examples are high-spatial frequency 'noise', usually widespread on the entire point cloud, or isolated points significantly differing from their neighbours (often also a product of measurement error, and typically referred to as 'outliers' because of the way they can be algorithmically detected, see section 2.3.2).

Several point cloud pre-processing operations are available to help compensate problems related to non-uniform coverage and measurement error. Some operations are dedicated to reducing noise or removing isolated points, while others are applied to improve the uniformity of sampling density (uniformity of point spacing). In the following sections, an overview of the most common pre-processing operations is presented.

2.3.1 Point cloud simplification, decimation and resampling

Point cloud simplification is the process of reducing the number of points in a cloud. The operation can be performed to reduce the computational complexity in the subsequent analysis operations, to reduce memory occupation, or to remove unwanted points, such as isolated points, overlapping points or points falling on surfaces other than the measurand ones (figure 2.9). An operation to reduce the overall number of points in a cloud is frequently performed at the beginning of the point cloud analysis process, given the high sampling densities commonly achieved by most coordinate measurement technologies.

Point cloud simplification may be performed by deletion of selected points, a process also known as decimation, or by replacing a set of existing points with a smaller-sized set of new points, usually obtained by some form of interpolation or other aggregation mechanism starting from the original set. The replacement operation

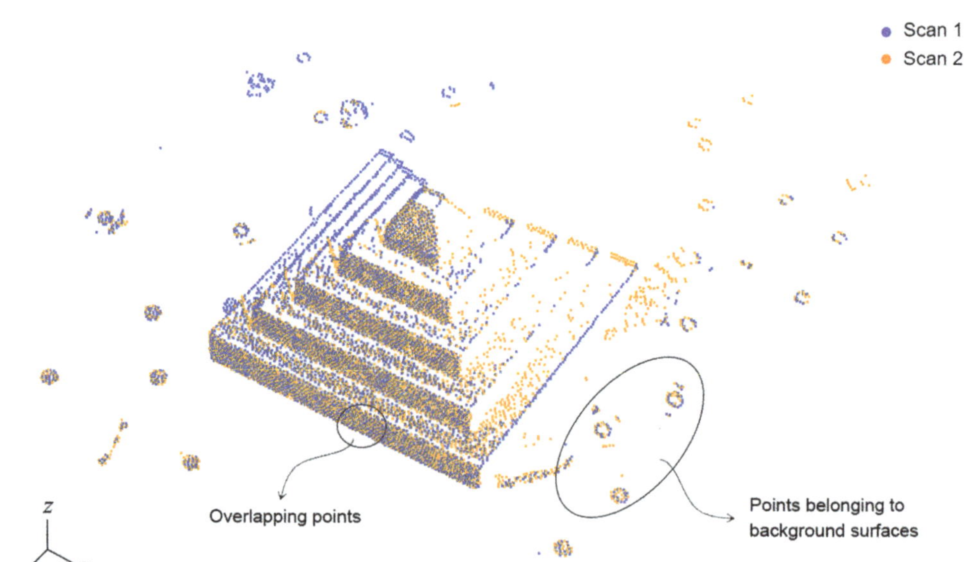

Figure 2.9. Example scenario involving clouds with excess points and requiring simplification. Two point clouds (blue and orange) obtained from different measuring the same test artefact for two different directions of observation. The clouds have been registered (i.e. co-localised in space see section 2.6) so that they can be merged into a single dataset. Regions captured by both the point clouds feature excess points; additional, isolated clusters of points are also present, representative of background surfaces that have been accidentally captured.

is typically referred to as 'point cloud resampling', to evoke the idea that points are the result of sampling on a continuous surface. It is worth pointing out that resampling in general may not necessarily imply a reduction in size of the point cloud and it may also be found in contexts different to point cloud simplification. In the specific case of dataset simplification, resampling is applied when the goal is to reduce the total number of points while at the same time obtaining a more uniform sampling density, if a simple decimation process does not suffice.

In general, operations involving the reduction of size of a point cloud (either by decimation or by resampling) imply reduction of spatial density, with a consequent loss of detail. Particularly in relation to the smaller-scale geometric features, an aggressive reduction may lead to the loss of definition of the underlying geometry and, therefore, to the loss of quality in the inspection result (Schall *et al* 2008). An additional risk of simplification based on resampling is related to the introduction of interpolated (i.e. non-measured) points.

In the literature, a large variety of simplification, decimation and resampling methods are discussed. Simplification methods have been introduced, for example, by Linsen (2001) and Pauly *et al* (2002), who developed algorithms to create simplified point clouds as true subsets of the original point set, extending Taubin's (1995) Laplacian operator method (originally developed for triangle meshes) to point clouds. An overview of simplification methods is given by Xiaohui *et al* (2007), in which sampling methods are divided into three categories: resampling, iterative simplification and clustering. The resampling category includes methods for the computation of a new sampled point cloud, considering specific features from the original dataset as key factors, that lead to a reduction of the number of points. The iterative simplification collapses pairs of points into a single data point, preserving the density of the point cloud. Clustering methods are often used to reduce the complexity of 3D point clouds, subdividing the bounding box into grid cells and replacing the original points that fall into the same cell by a common representative (see previous section 2.2.3.3).

2.3.2 Elimination of isolated points and noise reduction

A point in a point cloud is identified as isolated when it is significantly distant from its neighbours. The assessment of significance for distance is typically based on some global or local threshold (see later in this section). Isolated points are classified as:

- isolated points due to distance measured orthogonal to a local surface fitted to the neighbours (i.e. isolated points away from the surface—see figure 2.10(a)); and
- isolated points that still approximately lie on the locally fitted surface, but are too distant from their neighbours, as measured on the surface itself (i.e. isolated points on the surface—see figure 2.10(b)).

Isolated points on the surface are perceived as related to a local decrease of density or to the presence of 'void' regions; they can be either eliminated or they can be filled by resampling, as previously discussed in section 2.3.1 (for example, by clustering based on cell decomposition). Isolated points detected away from the locally fitted

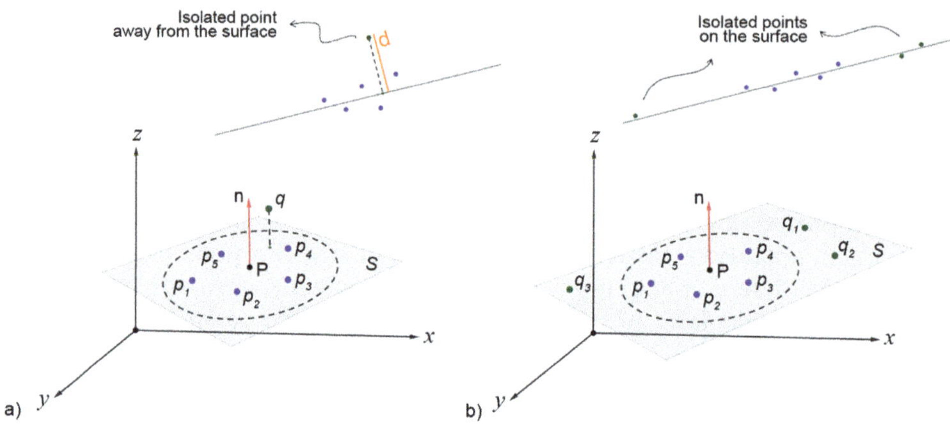

Figure 2.10. Types of isolated points: (a) isolated point away from the surface—isolation of point $\boldsymbol{q}$ is due to anomalous distance of the point to the fitted plane or surface S; (b) isolated point on the surface—isolation of point $\boldsymbol{q}$ is due to anomalous distance of the point to its closest neighbours, even though $\boldsymbol{q}$ approximately lies on the fitted plane or surface S.

surface may not necessarily imply a local decrease of density and are typically perceived as 'spikes' with respect to the surrounding points. Points away from the surface can be either eliminated, creating a local void, or replaced by points located closer to the surface. In most cases, the identification of a threshold of significance for distance is based on outlier statistics (Schall *et al* 2005, Sotoodeh 2006, Sultani and Ghani 2015). For example, the typical method for detecting whether the query point $\boldsymbol{q} \in \mathbf{P}$ is a 'spike' works as follows (Rabbani *et al* 2006, Pauling *et al* 2009):

(a) A surface (plane, low-order polynomial or spline) is locally fitted to a set of k points $\boldsymbol{p}_k$ selected as neighbours of $\boldsymbol{q}$. In an alternative to fitting, PCA can be used to determine the orientation of the local plane. In the fitting process, $\boldsymbol{q}$ is excluded.
(b) The local normal to the fitted plane or surface is computed, at the location identified by $\boldsymbol{q}$.
(c) The distances of all the $\boldsymbol{p}_k$ points to the fitted plane or surface are computed, measured along the direction of the local normal, and used to build a probability distribution.
(d) The distance of the query point $\boldsymbol{q}$ to the fitted surface, also computed along the local normal, is tested against the distribution of distances obtained for the neighbouring points, to see whether it must be considered as an inlier or outlier, using one of the many available outlier testing methods from statistical analysis.

For example, assuming that the distribution of distances can be modelled by a Gaussian function with a mean μ_k and a standard deviation σ_k, an outlier can be defined as any value whose difference from the mean is larger than a given multiple of the standard deviation (for example, $\mu_k \pm 3\sigma_k$). A similar formulation for outliers can be adopted to recognise isolated points on the surface, as long as the probability distribution is built on distances between neighbours on the fitted surface (i.e. not orthogonal to it). Finally, again, the same formulation can be adopted to detect isolated points in any direction, by considering the local probability distribution of distances considered in three dimensions.

Removal of noise from point clouds is typically implemented following methods similar to those illustrated for spikes. Noise is typically seen as high (spatial) frequency disturbances of point placement, measured orthogonally to the locally fitted surface and distributed over the cloud. Moving windows whose orientation follows the local normal can be implemented that, akin to signal processing, apply kernels based on mean or median filtering operations. As for spikes, local fitting can be replaced by local PCA depending on needs. Because of the parallelism with signal processing, noise removal and spikes removal operations are frequently referred to as point cloud 'filtering' operations. A comprehensive analysis of the state-of-the-art methods for filtering point clouds can be found in Han *et al* (2017).

2.3.3 Triangle meshes and conversion to/from point clouds

Another pre-processing operation which is often implemented for point cloud analysis is conversion to or from triangle meshes. A triangle mesh is a set of topologically connected triangles in a 3D space, typically defined by a set of vertices and connectivity between vertices (two connected vertices form an edge, three connected vertices—or three connected edges—form a triangle; vertices may be shared by multiple edges and triangles). Example of triangle mesh representations are shown in figure 2.11.

The conversion from a triangle mesh to a point cloud is typically trivial, as the vertices of the mesh can simply be converted into the points of the point cloud. The conversion from mesh to cloud is used when only vertices are needed for subsequent steps of the analysis, for example, when a specific property needs to be assessed which solely depends on points in a dataset.

However, the inverse task, i.e. converting a point cloud into a mesh, is challenging. At first glance, the points of a point cloud could be directly used as

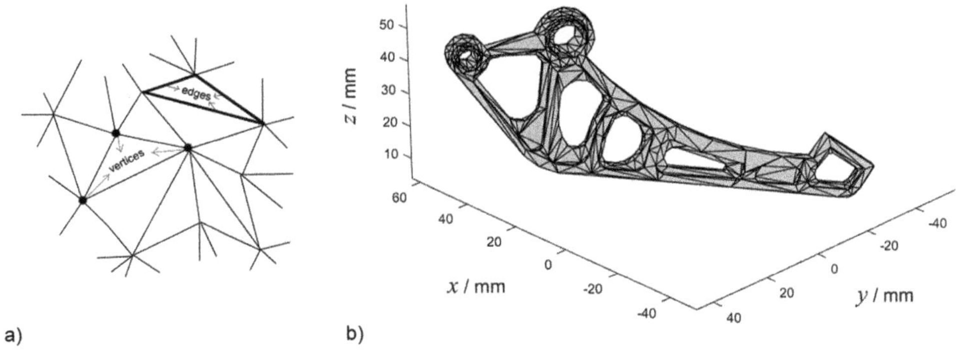

Figure 2.11. Example of triangle mesh representation: (a) detail of vertices, edges and triangles defining the mesh; (b) automotive part (roof bracket) represented as a triangle mesh (designed by the Centre for Additive Manufacturing, University of Nottingham).

vertices of the mesh. This is in principle correct. However, a triangle mesh also consists of the connectivity between vertices (to/from edges and then triangles), and the generation of a connectivity map is far from straightforward. Typically, in measurement, the points are considered as the result of sampling a continuous surface. The conversion to a triangle mesh (i.e. the generation of connectivity) would, therefore, somewhat restore the continuity of the surface from which the points were originally sampled. However, without any further knowledge of the measurand surface, there is no robust way of knowing whether two points really belong to the same portion of locally continuous surface, even if they appear very close to each other in space. For example, points sampled from two opposite surfaces of a very thin wall may still appear close to each other, yet they should not be considered as connected because they belong to different surfaces.

One of the most popular methods for reconstructing a triangle mesh from a point cloud is the application of Delaunay triangulation (Edelsbrunner 2009). In a 2D space, the Delaunay triangulation (DT) for a given set **P** of points is a triangulation DT(**P**), such that no point in **P** lies inside the circumcircle of any triangle in DT(**P**). The 2D form is preferred for point clouds that do not feature re-entrant parts and can, therefore, be projected onto a plane where the triangulation takes place. Once the 2D triangles have been generated, the 3D geometry can be reconstructed by adding back the coordinate along the direction of projection. A natively 3D form of the Delaunay triangulation also exists that can be applied to any point cloud; however, it generates tetrahedra, meaning that the final triangles must be extracted at a later step by identifying the outer faces of the tetrahedra, i.e. faces that are not shared by more than one tetrahedron (De Berg *et al* 2008). The general problem with Delaunay triangulation (2D or 3D) is that connectivity is established following criteria driven by the desire to obtain triangles that conform to specific geometric properties (no point of the original cloud may lie inside any given triangle or tetrahedron). Such geometric properties may not necessarily guarantee that the connectivity is correct for the underlying surface that the method is trying to reconstruct. Many other methods exist to establish connectivity and create a triangle mesh from a point cloud (Owen 1998, Bommes *et al* 2013, Berger *et al* 2014, 2017). The problem of converting a point cloud into a triangle mesh is a special case of a more general problem known as 'surface reconstruction', where the goal is to start from a point cloud and obtain a continuous surface expressed in some analytical form, from which the cloud might have been sampled. The problem of surface reconstruction can be addressed considering any type of continuous geometry, including polynomial surfaces and parametric surfaces (splines, b-splines, non-uniform rational basis splines NURBS). Reconstruction into a triangle mesh is, therefore, seen as a special case of reconstruction where the surface is defined in a very simple way (a set of connected triangles). The conversion of point clouds into triangle meshes is of primary importance, given the large diffusion of triangle meshes in many applications dealing with 3D geometries (Kobbelt and Botsch 2004). In the context of coordinate metrology, and the themes covered by this chapter, evidently the possibility of reconstructing a surface in the form of a triangle mesh may provide significant benefit to all the other pre-processing and point cloud analysis methods discussed. This is primarily because a triangle mesh allows for a significantly more robust selection of the true neighbours of any given query point, whilst without connectivity, only selecting points based on distances cannot be relied upon.

2.4 Point features and partitioning

In point cloud analysis, a cloud feature is a property/attribute that describes the entire cloud. The dimensions of an AABB encompassing the whole cloud are examples of cloud features. The centroid and the aspect ratio between the principal axes computed on the point cloud are other examples of cloud features. Conversely, a point feature is any property/attribute that can be associated to each individual point of a point cloud. Examples of point features may be colour, local orientation or local curvature. The position of a point is a point feature itself, and it is the only feature that must be mandatorily defined for every point in the point cloud. Unless point features are directly available from measurement (for example, colour information, if the measuring system provides the option to capture it), most point features must be computed, and necessitate point neighbour information in order for the computation to take place. The local normal, for example, needs knowledge of point neighbours in order to identify a local, reference plane (see next section 2.4.1). Local curvature also needs neighbour information (see section 2.4.2). Because of the need for neighbours, many point features essentially capture properties related to the local 'shape' of the surface (as approximated by the point itself and its immediate neighbours) and are, therefore, often referred to as local shape descriptors. A survey on point features for 3D point clouds is reported in Han *et al* (2018).

Cloud features and associated similarity/difference metrics can be used to compare point clouds, recognising those that are similar to each other, or different based on size, position, aspect ratio, etc. Analogously, point features and associated similarity/difference metrics can be used to compare points, recognising those that are more similar or different from each other, based on any attribute that can be computed as a point feature.

2.4.1 Point normals

The estimation of the local normal to a query point $\boldsymbol{q}$ (figure 2.12(a)) is usually performed by fitting a plane T (a low-order polynomial surface or a spline) to $\boldsymbol{q}$ and its closest neighbours $\boldsymbol{p}_k$, found using the k-NN algorithm. Then, the point normal $\boldsymbol{n}$ is identified by the normal to the fitted plane (Hoppe *et al* 1992).

Alternatively (figure 2.12(b)), once the local subsets of neighbouring points have been selected via the k-NN algorithm, PCA can be performed (Pauly *et al* 2002, Holzer *et al* 2012).

The PCA is computed on the union of the query point $\boldsymbol{q}$ and its $\boldsymbol{p}_k$ neighbours (see section 2.2.2 for PCA general formulation on entire dataset $\mathbf{P}$). The covariance matrix $\mathbf{C}$ shows how subsets of neighbouring points locally vary from their centroids $\boldsymbol{c}$. Given v_j the eigenvectors and λ_j, the corresponding eigenvalues and $j \in \{0, 1, 2\}$

$$\mathbf{C} \cdot v_j = \lambda_j \cdot v_j. \tag{2.9}$$

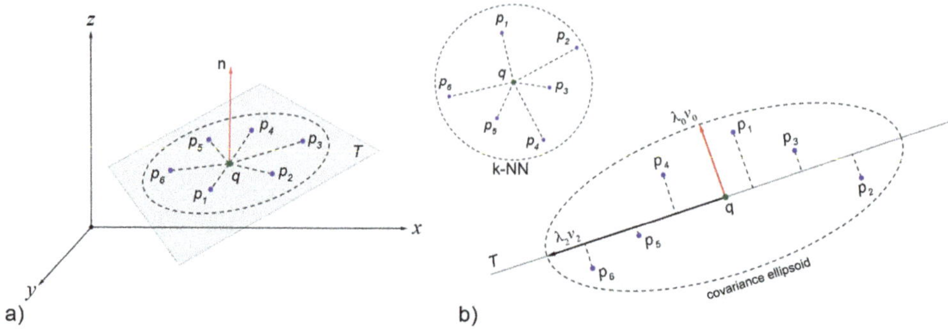

Figure 2.12. Point normal estimation approaches: (a) fitted plane T to a query point $\boldsymbol{q}$ and its neighbours $\boldsymbol{p}_k$; $\boldsymbol{n}$ to the point $\boldsymbol{q}$ is identified by the normal vector to the fitted plane. (b) Once the neighbouring points are found using k-NN, the normal estimation can be found via PCA; given the covariance matrix built with the coordinates of local points, the eigenvector v_0 with the smallest eigenvalue λ_0 corresponds to the normal vector $\boldsymbol{n}$.

If $\lambda_0 \leqslant \lambda_1 \leqslant \lambda_2$, the eigenvector v_0 with the smallest eigenvalue λ_0 identifies the normal vector to the fitted tangent plane T, geometrically perpendicular to the eigenvector v_2 with the greatest eigenvalue λ_2, in the direction of the principal variation.

In many operations for point cloud processing, point normals are required to be consistently oriented—the normals must all be directed inwards or outwards with respect to a hypothetical surface, which ideally encloses the geometric object from which the points have been sampled. As a convention, normals associated to the surface of a geometric object are typically assumed as pointing outwards (figure 2.13). However, the methods illustrated above for computing point normals do not guarantee consistent orientation, thus some type of post-processing operation must be typically applied.

A method to compute a consistent orientation of the normal vectors is based on using Euclidean minimum spanning trees (EMST, see section 2.2.3.4). Starting from a subset $\mathbf{Q} \subset \mathbf{P}$ of points whose normals are assumed as correctly oriented, EMST is used to reach neighbours located within a region of increasing radius. The normal $\boldsymbol{n}_j$ of each new neighbour $\boldsymbol{p}_j$ is checked against the (correct) normal $\boldsymbol{n}_i$ of the previous (correct) point $\boldsymbol{p}_i$, and flipped if inconsistent using

$$\boldsymbol{n}_i \cdot \boldsymbol{n}_j < 0 \rightarrow \boldsymbol{n}_j = -\boldsymbol{n}_i. \tag{2.10}$$

2.4.2 Point curvatures

In differential geometry, k_1 and k_2 are the maximum and minimum local curvatures of the planar curves formed by intersecting a normal plane with the surface itself (Fisher 1989). The principal curvatures can be used as local shape descriptors (Bär 2010, Pressley 2010).

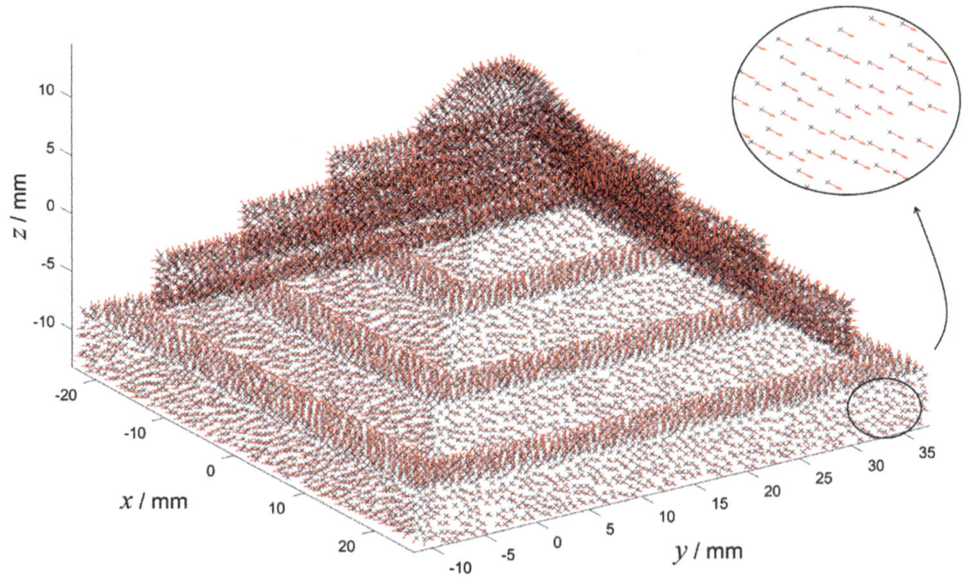

Figure 2.13. Point vector normals consistently oriented pointing outwards, shown on pyramid artefact designed by Manufacturing Metrology Team, University of Nottingham.

Surface curvature estimation has been studied extensively over the past decades, although mostly dedicated to continuous surfaces. For point clouds, methods for computing curvatures rely on locally fitting the point and its neighbours to a continuous surface onto which the curvatures are computed, or rely on using a triangle mesh as a substitute for the continuous surface (Meyer *et al* 2003, Magid *et al* 2007). More recently, methods have been proposed for curvature computation directly on point clouds (Mérigot *et al* 2011). An example method consists of running a PCA on the local normals (i.e. not on the point coordinates) belonging to the neighbours of $\boldsymbol{q}$, and estimating the principal curvatures using the eigenvectors and eigenvalues of the resulting covariance matrix.

Two additional types of curvature, often computed on point clouds, are the Gaussian curvature K and the mean curvature H (Dyn *et al* 2001). These can be estimated from the principal curvatures k_1 and k_2 as follows

$$K = k_1 \cdot k_2, \tag{2.11}$$

$$H = \frac{(k_1 + k_2)}{2}. \tag{2.12}$$

In figure 2.14, mean and Gaussian curvature values are mapped to a colour scale for visualisation purposes. Mean and Gaussian curvature can be used as local shape descriptors, for the purpose of segmentation, classification, comparison of shapes and registration (see further discussion in section 2.6).

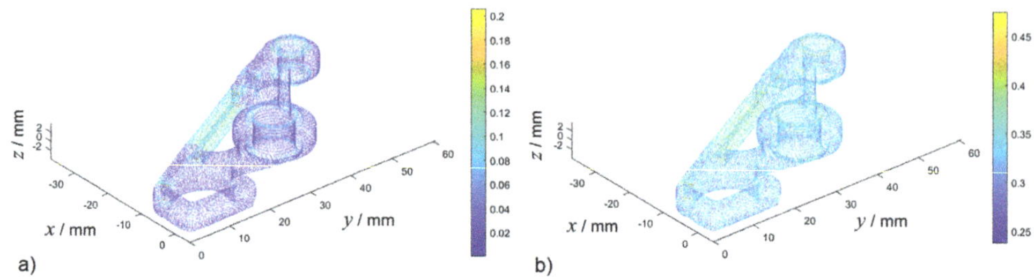

Figure 2.14. Local curvature values estimation on point cloud (roof bracket), converted into a colour map representation: (a) Gaussian curvature K and (b) mean curvature H.

2.4.3 Partitioning and segmentation

By using point features, for example normals and curvature values, as illustrated in section 2.4.1 and section 2.4.2, points can be classified, i.e. mapped to a discrete and countable number of classes. Classification can be implemented in multiple ways:

(a) by threshold-based rules (by explicitly setting a maximum degree of similarity for point features to belong to the same class);
(b) by statistical clustering, i.e. automated identification of thresholds to produce groups of similar point features; and
(c) by region growing (a combination of thresholding and clustering, starting from points of known classification and expanding the result via incorporation of consistent neighbours).

Regardless of the classification method, any classification operation results in a partitioning, i.e. generation of subsets of points, corresponding to the identified classes. If the partitioning operation leads to subsets containing points that are also topologically adjacent (assuming that the topological relationships can be reconstructed—see section 2.3.3), then such subsets are called segments. In parallel with digital image processing, a segment is in fact a set of adjacent pixels classified as belonging to the same class (Szeliski 2011). A classification operation on point features leading to segments is called a point cloud segmentation (Nguyen and Le 2013).

Point cloud segmentation paves the way for shape representations that rely on decomposition, favouring the implementation of shape-based automated reasoning solutions powered by machine learning (Catalucci *et al* 2019) and supporting applications where semantic interpretation of regions is needed (for example, decomposition of a teapot model to identify its handle). Reviews of the most common segmentation methods for point clouds can be found elsewhere (Rusu and Cousins 2011, Nguyen and Le 2013, Grilli *et al* 2017, Xie *et al* 2020).

2.4.3.1 Partitioning of point clouds based on k-means and hierarchical clustering

Clustering based approaches (for example, *k*-means and hierarchical clustering) applied to point clouds generate partitions/clusters where each cluster gathers points with similar point feature values. Similarity/difference can be computed in multiple ways, for example by using Euclidean or city-block distance, depending on the dimensionality of the feature (for example, a scalar or a n-dimensional vector) (James *et al* 2013).

In point cloud k-means clustering, a final number of clusters (k) must be decided in advance. The method starts with an initial random assignment of points to clusters, then it proceeds by iteratively refining the assignments using considerations based on similarity/difference. Clusters are updated at each iteration, so that point assignment evolves dynamically until the last point has been classified. An example application of k-means clustering is shown in figure 2.15(a), where points are assigned to one of five clusters ($k = 5$) based on Gaussian curvature. Note how points belonging to the same cluster may not necessarily be adjacent to each other; that is, partitioning by k-means clustering is not a segmentation method.

Hierarchical clustering of points in point clouds is an alternative to k-means that does not involve the specification of the number of clusters needed. Instead, the most similar points (where similarity is computed on point features) are grouped into proto-clusters, then iteratively aggregated into larger ones, following between-cluster similarity measures. The iterative aggregation process is described as a tree-like graph known as 'dendrogram', which invariably terminates with a single cluster incorporating all points of the point cloud. Criteria can be set to stop the recursive aggregation once specified criteria are met (for example, the maximum difference between clusters that still allows their aggregation). An example application of hierarchical clustering for point cloud partitioning is shown in figure 2.15(b), again using Gaussian curvature as the point feature driving the classification. As for k-means, the partitioning result is not a segmentation, because points may belong to the same cluster even if they are far apart from each other. Clustering is often used to isolate subsets of points whose point features are extremal in value, for example, to isolate points with the highest curvature values. Such points may act as landmarks for subsequent registration processes, as illustrated in section 2.6.1.

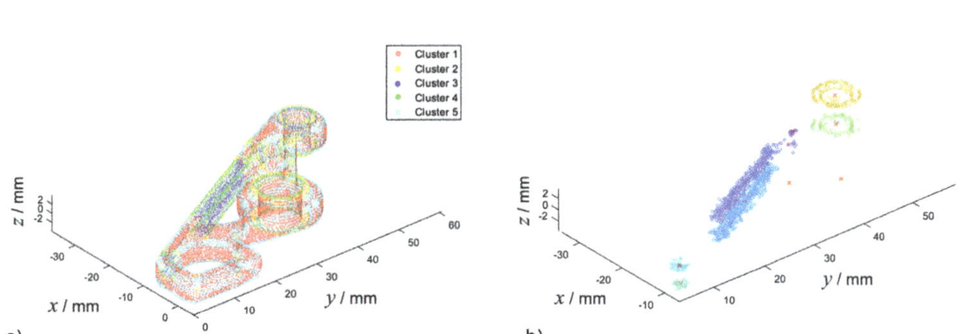

Figure 2.15. Point cloud partitioning methods. (a) k-means clustering on Gaussian curvature (points are assigned to one of five clusters ($k = 5$)); cluster 3 refers to the points with the highest curvature values. (b) Hierarchical clustering on Gaussian curvature and cluster centroids. The points taken into account are the ones with the highest curvature values.

2.5 Point cloud fitting

Point cloud fitting is defined as the process of identifying a geometric surface model that best describes the point cloud (Forbes and Minh 2011, 2012). In layman's terms, fitting means finding a geometric model whose surfaces are sufficiently close to the points of a set to support the assumption that such points might have been sampled from that geometry.

Point cloud fitting is commonly used for a variety of reasons, such as replacing a point cloud with a continuous, mathematically defined geometry, investigating local properties of the point cloud by approximating a local cloud subset to a more 'readable' geometric entity (for example, fit a local subset of points to a plane to obtain the local normal), obtaining reference geometric datums to perform further computation of distances, lengths, etc, and, finally, for comparing a point cloud (assumed from measurement) to a nominal reference (the fitted geometric model).

A point cloud fitting process is comprised of two steps:

(a) choosing a geometric model to which to fit the point cloud; and
(b) choosing a fitting method.

The geometric model to fit to is called a 'substitute feature'. The choice of substitute feature influences the type and number of geometric modifications that the fitting process can apply to the feature itself, in order to make it a better fit to the point cloud; for example, a parametric, cylindrical surface could be altered by operating on the length and diameter (parameters) to obtain a cylinder that better fits a given point cloud. A polynomial surface may be fit to the point cloud by changing the orders of the underlying polynomials. The substitute feature may also be a triangle mesh. In this latter case though, the transformations imposed by fitting may be limited to a change of scale, or more frequently to a single rigid transform (translation and rotation), which turns the fitting problem into a registration problem (Tam *et al* 2013)—see section 2.6. In metrology, fitting to a triangle mesh is commonly used to perform a coarse comparison between a nominal geometry (the triangle mesh) and the result of a measurement (the point cloud). More challenging is the choice of a substitute feature when fitting involves clouds sampled to freeform or otherwise complex surfaces. This is a frequent issue with additively manufactured parts (Forbes and Minh 2011).

The second step of fitting is the choice of the mathematical process through which the chosen analytical geometric model (substitute feature) is adapted to the point cloud. In general terms, fitting is usually implemented as some type of minimisation involving the distances of the points to the best-fit geometric model. In the following sections, the most common fitting methods are presented.

2.5.1 Fitting methods

The geometric model (substitute feature) can be assumed as a mathematically defined, continuous surface. In general, this can be expressed by the parametric function $f(\boldsymbol{u}; \boldsymbol{b})$ where $\boldsymbol{b}$ is known as the shape vector (Feng and Hopp 1991) which changes depending on the type of mathematical representation adopted, and $\boldsymbol{u}$ is a vector of parameters to navigate across the continuous surface (a generalisation of surface coordinates). Most fitting methods are based on the L_p norm, given by

$$L_p \doteq \left[\frac{1}{N} \sum_{i=1}^{N} | e_i |^p \right]^{1/p}, \tag{2.13}$$

where N is the number of points in the cloud and e_i is the shortest Euclidean distance between the ith point of the cloud and the nearest point on the continuous surface defined by the substitute feature. The best-fit feature is the one that minimises L_p. Since this is a minimisation problem, $1/N$ and the exponent $1/p$ can be omitted, leading to the following expression to be minimised

$$E_p \doteq \sum_{i=1}^{N} | e_i |^p, \tag{2.14}$$

where E_p is a measure of cumulative 'error' between the point cloud and the substitute feature.

2.5.1.1 Least-sum-of-distances fitting

When $p = 1$, equation (2.14) becomes

$$E_1 = \sum_{i=1}^{N} | e_i |, \tag{2.15}$$

where E_1 is a sum of absolute distances. The minimisation of E_1 is called least-sum-of-distances fitting. The minimisation problem can be solved using a number of optimisation methods (Feng and Hopp 1991).

2.5.1.2 Total-least-squares fitting

When $p = 2$, equation (2.14) becomes

$$E_2 = \sum_{i=1}^{N} | e_i |^2 = \sum_{i=1}^{N} e_i^2, \tag{2.16}$$

where E_2 is called the sum of squared distances, and the minimisation problem is known as total-least-squares (TLS) fitting, also referred to as orthogonal distance regression (Gruen and Akca 2005, Monserrat and Crosetto 2008). Note that, as stated previously, e_i is the Euclidean distance between point $\boldsymbol{p}_i$ and the corresponding closest point on the geometric model. Assuming that such a closest point can be found using the value $\boldsymbol{u}_i$ for the parameter vector $\boldsymbol{u}$, the minimisation of the sum of squared distances is expressed as

$$\min_{b} \left(\sum_{i=1}^{N} \left| \boldsymbol{p}_i - f(\boldsymbol{u}_i; \boldsymbol{b}) \right|^2 \right). \tag{2.17}$$

Equation (2.17) describes the minimisation problem as the problem of finding the optimal values for the components of the shape vector $\boldsymbol{b}$ so that E_2 is minimised. The minimisation problem is in general non-linear and must be solved via iterative methods (for instance, Gauss–Newton or Levenberg–Marquardt), although the risk of converging to local minima must be taken into account.

In general, TLS fitting is the most frequently applied fitting method for point clouds. However, least-sum-of-distances fitting is also applied, in particular as it is slightly more robust to outlier points.

2.5.1.3 Ordinary-least-squares fitting

In some cases, when the nature of the point cloud allows it, fitting could be done to a function of type: $z = f(x, y; \boldsymbol{b})$. In this formulation, the substitute feature is described by a dependent variable (z) function of the variables x, y, which define the location and essentially replace the vector of parameters $\boldsymbol{u}$, and $\boldsymbol{b}$ is still the shape vector of the surface, the usual independent variable in the minimisation problem. When this formulation is possible, the L_p norm can be modified by defining the Euclidean distance between each point and its closest neighbour in the substitute feature (e_i) as a simple difference in z coordinates. For the $p = 2$ case (least-squares), the minimisation problem is expressed as

$$\min_{b}\left(\sum_{i=1}^{N}\left| z_i - f\left(x_i, y_i; \boldsymbol{b}\right) \right|^2\right), \tag{2.18}$$

which is an example of ordinary-least-squares (OLS) formulation as opposed to TLS. The OLS problem is computationally simpler, and in some cases, it admits closed-form solution. For example, when considering the OLS fitting of a point cloud to a plane, equation (2.18) becomes

$$\min_{b}\left(\sum_{i=1}^{N}\left| z_i - \left(Ax_i + By_i + C\right) \right|^2\right), \tag{2.19}$$

where parameters A, B and C are the components of the shape vector $\boldsymbol{b}$ that must be found in order for the expression to be minimised. Turning a point cloud fitting problem into the OLS problem defined by equation (2.19) is appealing as the formulation admits closed-form solution. However, the field of applicability of equation (2.19) is limited by the requirement of the point cloud being representable by $z = f(x_i, y_i; \boldsymbol{b})$, which automatically excludes re-entrant surfaces or full-3D geometries.

The difference between solving the OLS problem or the TLS problem is visually exemplified in figure 2.16, where a 2D point cloud is fit to a line.

2.5.1.4 Two-sided and one-sided minmax fitting

Two-sided minmax fitting consists of the minimisation of the L_p norm when $p = \infty$.

When $p = \infty$, it can be demonstrated that

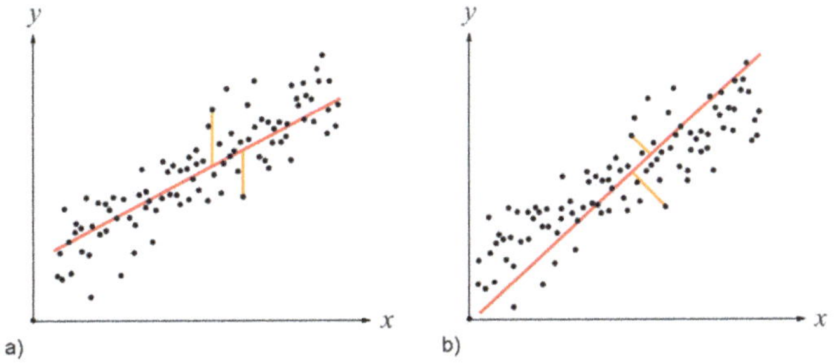

Figure 2.16. 2D representation of point cloud fitted to a line. Difference between (a) ordinary-least-squares (OLS) fitting and (b) total-least-squares (TLS) fitting.

$$L_\infty = \lim_{p\to\infty} L_p = \max_{1\leqslant i\leqslant N} \mid e_i \mid. \tag{2.20}$$

In other words, L_∞ is the maximum distance of the points in the cloud computed on both sides of the geometric surface model. The minimisation of L_∞ is expressed as

$$\min_b \left(\max_{1\leqslant i\leqslant N} \mid e_i \mid \right), \tag{2.21}$$

which means that at the end of the fitting process, the maximum signed e_i will be reduced to its minimum possible value, or in other words, that the maximum signed distance of the sampled data points and the geometric model is minimal on both sides. Because of this effect, the minimisation is called 'two-sided minmax fitting'.

With two-sided minmax fitting, the fitted model will lie approximately in the centre of the data points, in a visually similar manner to the results of a TLS fitting (though the two methods solve slightly different minimisation problems). In some circumstances, however, a fitting problem must be solved to identify a substitute feature that lays entirely on one side of the point cloud. A typical example is fitting to the minimum circumscribed circle or to the maximum inscribed circle. Another example is the identification of the lowest plane 'touching' the cloud from above, and the highest plane touching it from below. Geometric tolerances are typically assessed by solving these types of problem. Minmax fitting can be used to solve these problems as long as the minimisation is accompanied by a constraint on the sign of the errors (e_i) that can be involved in the minimisation. For instance, if only the minimisation of the positive e_i is considered

$$\min_b \left(\max_{1\leqslant i\leqslant N} \mid e_i \mid \right), \quad e_i \geqslant 0, \forall\, i = 1, \cdots, N. \tag{2.22}$$

A two-sized minmax fitting, where minimisation is constrained on the sign of e_i, is called 'one-sided minmax fitting'. Note that one-side minmax fitting is not the only method available to solve one-sided fitting problems and to identify datums for tolerances, for example see Moroni and Petrò (2008) for a survey on minimum-zone fitting.

2.5.1.5 RANSAC fitting

Random sample consensus (RANSAC) is a general framework for model fitting in the presence of a large number of outliers. Introduced by Fischler and Bolles (1981), RANSAC is a robust iterative algorithm used to estimate parameters of mathematical features (for example lines or circles) from a set of observations that contains outliers. The name comes from the actual algorithm functionality of randomly selecting a sampled group of data points in datasets containing both inliers and outliers, and being able to consider only the points classified as inliers to perform the optimal fitting. The set of inliers obtained for the fitting is called the consensus set. RANSAC algorithms achieve their goals by iteratively repeating the steps that follow (Fischler and Bolles 1981):

- Select a random subset of N points of the original dataset $\mathbf{P}$. This subset is identified as a group of hypothetical inliers.
- Fit a model x to the set of hypothetical inliers.
- Test the rest of the data against the fitted model; the points that fit the estimated model well are considered as part of the consensus set K, within a user-given threshold.
- If a sufficient number of points is classified as part of the consensus set, the estimated model x is reasonably good and the process terminates; otherwise start a new iteration with another random subset.

The iterative procedure is repeated until the consensus set is large enough, and/or until a predefined number of iterations is reached. Reaching the latter condition may mean that a proper fitting has not been obtained. However, the advantages with respect to other methods, such as total or ordinary-least-squares fitting, is that with RANSAC fitting it is possible to take outliers into better account, whilst in least-squares methods, all points are accounted for in the same manner (unless weighted formulations are adopted).

Because of its intrinsic robustness to outliers, RANSAC is often applied to solve point cloud-related problems other than fitting. For example, RANSAC is used as an outlier detection method (Strutz 2016), for automatic identification and extraction of shape primitives (Schnabel *et al* 2007) and for registration (Tam *et al* 2013, Catalucci *et al* 2019) (see also section 2.6.4).

2.6 Registration of point clouds

A point cloud produced by measurement is defined within a local coordinate system created by the measurement instrument itself. When multiple measurements are taken from the same object, for example from different viewpoints or using different instruments, each point cloud has its own coordinate system and in principle there is no explicit information that indicates how point clouds are spatially related to each other. Registration is the process that brings multiple point clouds taken from observations of the same scene in their correct, relative position within a shared coordinate system.

From a mathematical standpoint, registration implies the application of a transformation to each point cloud consisting of a rotation and a translation (see previous section 2.1.3). Such a transformation is called 'rigid' as it does not alter the shape of each point cloud (only its position and orientation) and the registration process is, therefore, called 'rigid registration'. There are also registration processes where transformations are allowed to slightly alter the shape of the cloud, for example by changing its scale, or by deforming the cloud to compensate for specific types of aberration introduced by measurement. When this happens, the registration process is called 'non-rigid registration'. In coordinate metrology, non-rigid registration is particularly challenging, because it could introduce additional errors to the already present measurement errors. In this section, rigid registration strategies will be the primary focus.

In optical measurement, rigid registration allows for multiple high-density point clouds to be aggregated from measurements taken at different viewpoints, in order to increase the part coverage. This process is known as point cloud stitching (figure 2.17).

In coordinate metrology, rigid registration can also be used to localise multiple point clouds taken from the same surface region, so that later aggregation can be used to reduce the random component of measurement error. Registration of point clouds acquired using multiple instruments can be used to compare the performances of the instruments themselves. Finally, registration of point clouds obtained by measuring different parts can be used to study geometric differences between the parts.

Registration may also be applied to localise triangle meshes, continuous surface representations and any hybrid combination of the previous types. Registration of

Figure 2.17. An example of point cloud stitching. High-density point sets acquired with an optical system from multiple viewpoints are registered together, in order to maximise the part coverage.

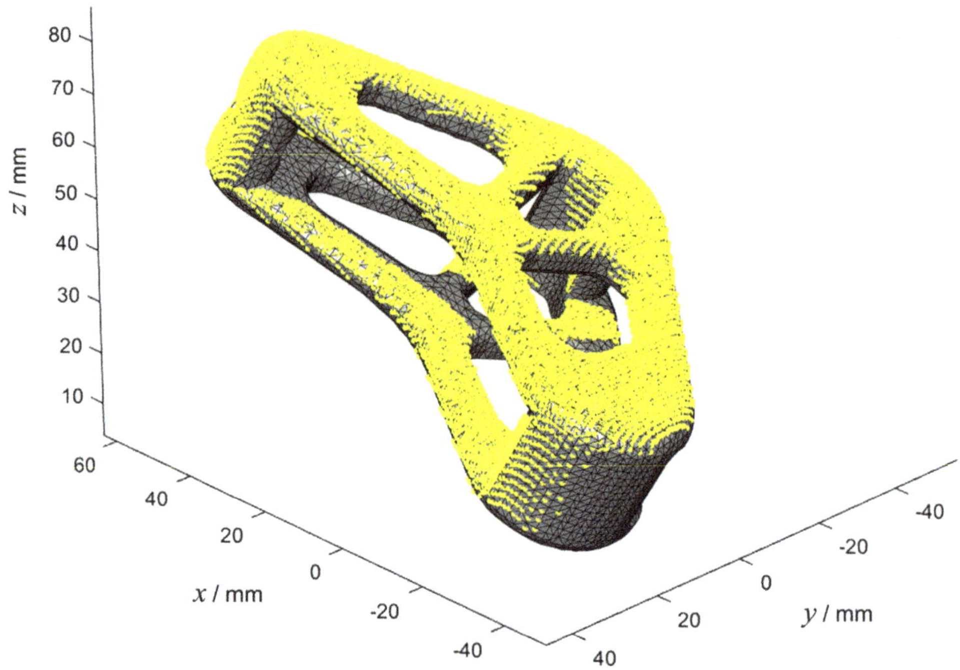

Figure 2.18. Example of registration between a point cloud to a continuous surface (i.e. triangle mesh). The point cloud simulates a unidirectional single shot optical measurement of an industrial part.

point clouds to continuous surfaces or triangle meshes in particular can be used to compare a measured geometry to a nominal geometry (figure 2.18), or to find a mathematical geometry that is representative of the cloud itself and can be used in substitution of the cloud. In this latter scenario, registration becomes a type of fitting (discussed in section 2.5). Rigid registration is equivalent to a fitting process where only rigid transformations are allowed.

Non-rigid registration methods can be adopted to tackle some fitting problems that involve modification of parameters of the substitute feature (for example, the scale). However, non-rigid registration methods in general do not cover an equivalently wide set of cases as other more general fitting methods (see section 2.5.1).

Finally, registration processes may be defined where localisation involves more heterogeneous datasets, for instance 3D point clouds with digital 2D images, or 3D point clouds with 3D voxel data. These types of registration processes are emerging in coordinate metrology but will not be covered in this section. The methods illustrated in the following always presume that at least one dataset to be registered is a point cloud, ideally but not necessarily produced by measurement.

An overview of most of the registration methods illustrated in this section can be found elsewhere (Tam *et al* 2007). Most registration methods only operate on two datasets at a time (point clouds or meshes); one typically being left in its original state while the other is transformed to achieve localisation in the coordinate system of the first. When multiple datasets must be registered, methods operating on pairs must proceed sequentially, either by registering each new dataset to the same reference one, or by registering each new dataset to the merged dataset obtained at the previous registration step, therefore, chaining new datasets in sequence. The chaining approach carries a larger risk of error accumulation, particularly when attempting to 'close the loop' on point clouds chain-stitched all around a measured object. Finally, as opposed to proceeding with pairs, multi-view registration methods are capable of processing all the datasets simultaneously. Further details on registration methods based on pairs or operating on all datasets at the same time can be found elsewhere (Pulli 1999, Gruen and Akca 2005, Brown and Rusinkiewicz 2007, Salvi *et al* 2007, Ramasamy *et al* 2013, Wang *et al* 2015).

2.6.1 Registration based on external references or based on matching landmarks

When either the part being measured or the measurement system can be mounted onto a robotic arm, contact coordinate measuring systems (CMSs) or other moveable indexed devices (for example, a part placed on a rotary stage), relative positioning can be recorded from encoders and can be used to perform the registration of multiple point clouds taken within the same set-up. The accuracy and precision of measurement is usually superior to that of the positioning system, thus registration results should usually be subjected to further refinement, using methods described later in this section.

However, information on relative positioning is often not present or may be inadequate. For example, in additive manufacturing, the mass customisation typical of layer-based processes often does not make it feasible or economically convenient to prepare dedicated fixtures, so even if using a rotary stage or other controlled positioning axes, small displacements between the part and the measurement point may occur.

When information on relative positioning is absent or not sufficiently reliable, the only option is to search for 'hints' in the point clouds themselves. For example, if measurements are taken from slightly different viewpoints but cover at least in part the same portion of the surface, region similarities between point clouds may be searched for as potentially representative of sampling from the same area. The approach of searching for similarities is known as the 'search for correspondences', or 'search for corresponding landmarks'. Once corresponding landmarks are found, the point clouds can be rotated and translated until the landmarks are overlaid, providing a best estimate of the registration result.

With the search for corresponding landmarks, the registration problem is turned into a problem of landmark identification, followed by their geometric overlay, the latter process often being referred to as geometric alignment. Geometric alignment of landmarks implies minimisation of their relative distances, which mathematically is not dissimilar to the problems dealt with by fitting. The main limitation of registration based on aligning matching landmarks is the impossibility to apply this approach to point clouds which do not contain regions sampled from the same surface.

2.6.2 The absolute orientation problem

At the heart of many contemporary registration methods based on landmark alignment lies the absolute orientation problem. The absolute orientation problem has a closed-form solution and is, therefore, extremely fast, not requiring iterations, and is guaranteed to deliver an optimal solution. In the absolute orientation problem (Horn 1987, Lourakis 2016), two point clouds are considered, one moving $\mathbf{P} = \{p_1, p_2, \dots, p_n\}$ and the other fixed $\mathbf{Q} = \{q_1, q_2, \dots, q_m\}$, not necessarily containing the same number of points ($m \neq n$). A number k of pairwise correspondences has been predefined; those correspondences are known as pairwise matches, involving k points per cloud ($k < \min(m, n)$). In the context of landmark-based registration, individual points act as landmarks and pairwise matches identify the unique aligned correspondences. Solving the absolute orientation problem means finding a transformation to be applied to the moving point cloud so that the Euclidean distances between pairwise matched points are minimised. The general transformation for **P** includes a rotation matrix **R**, a translation vector $\boldsymbol{t}$ and a global scaling coefficient s, that is

$$\min_{s,\mathbf{R},\boldsymbol{t}} \left(\frac{1}{k} \sum_{i=1}^{k} \| \boldsymbol{q}_i - (s\mathbf{R}\boldsymbol{p}_i + \boldsymbol{t}) \|^2 \right), \tag{2.23}$$

where $\boldsymbol{p}_i$, $\boldsymbol{q}_i$ is the pair of points belonging to the ith pairwise match. A simplified formulation that removes the scaling coefficient ($s = 1$) is commonly adopted, so that the absolute orientation problem can be embedded in rigid registration approaches, i.e.

$$\min_{s,\mathbf{R},\boldsymbol{t}} \left(\frac{1}{k} \sum_{i=1}^{k} \| \boldsymbol{q}_i - (\mathbf{R}\boldsymbol{p}_i + \boldsymbol{t}) \|^2 \right). \tag{2.24}$$

The above minimisation problem has a closed-form solution which has been found following multiple approaches (Arun *et al* 1987, Horn 1987, Horn *et al* 1988). Resolution methods are compared and discussed elsewhere (Eggert *et al* 1997). An analogous problem has been solved in statistical shape analysis under the name of 'Procrustes analysis' (Kendall 1989).

2.6.3 Alignment by means of principal component analysis

PCA (see section 2.2.2) can be used to perform alignment between multiple datasets. In this case, landmarks are not identified as subsets of existing points from the original point clouds; instead they are new geometric entities automatically generated by the PCA on each point cloud, namely the centroid and principal axes. Alignment by PCA (Jolliffe 2002) is always a rigid transformation unless the corresponding principal axes are scaled, in order to obtain a match in length. The rigid version of the procedure is tackled as follows (figure 2.19):

- the centroids of the datasets are overlaid by translation applied to the moving point cloud **P**; and
- a rotation is applied in order to align the first, second and third principal axes of the datasets.

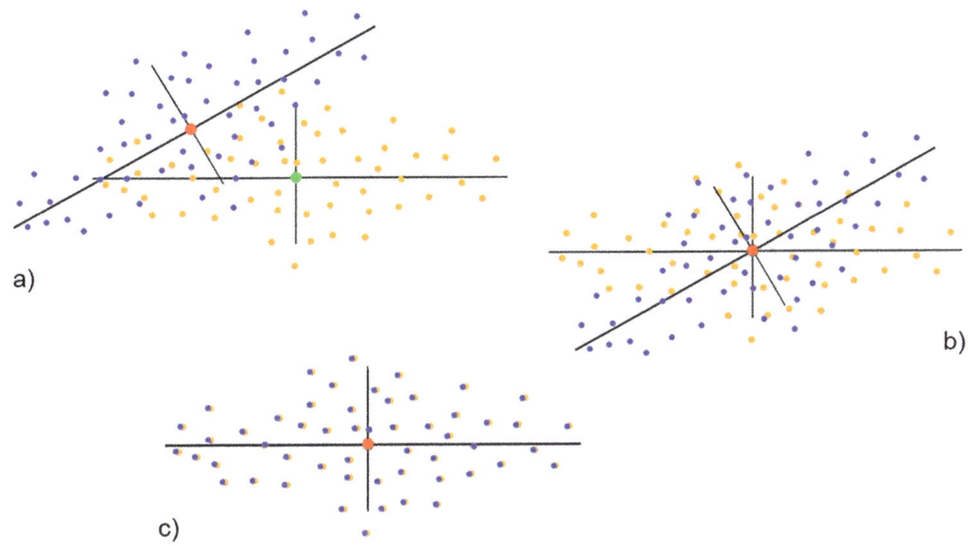

Figure 2.19. Point clouds alignment via PCA. (a) Two sets of points (blue and orange, respectively) in their initial configuration. (b) The centroid (red) of the first point cloud is translated in order to coincide with the centroid (green) of the second point cloud. (c) A rotation is applied in order to align the first, second (and third if 3D representation) principal axes of the datasets.

Alignment performed via PCA can always be applied to any pair of point clouds, even when they do not feature any apparent similarity (Salvi *et al* 2007). The downsides are related to the intrinsic properties of PCA—if the principal axes are very similar in length (i.e. have similar eigenvalues), the wrong axes may be paired. Even when the pairing is correct, there are two valid solutions for each axis (either the positive directions along the axis are matched or a positive is matched to a negative) leading to 180° rotations that are equally valid PCA results (figure 2.20).

The use of alignment by PCA is, therefore, confined to obtaining very crude initial approximated localisation of point clouds in space, to pave the way for the subsequent application of better performing methods.

2.6.4 RANSAC alignment

Recalling the absolute orientation problem as the core method to mathematically solve the alignment problem (see section 2.6.2), the challenge remains for the identification of correspondences between points (best pairwise matches), needed to solve the absolute orientation problem. Two solutions can be adopted:

1. The best matching points are those that can be better aligned, regardless of any other consideration about the nature of the selected points.
2. The best matching points are those that are the most similar, according to some metric defined using the points themselves and (usually) their immediate neighbours, regardless of whether such matches can be optimally aligned or not.

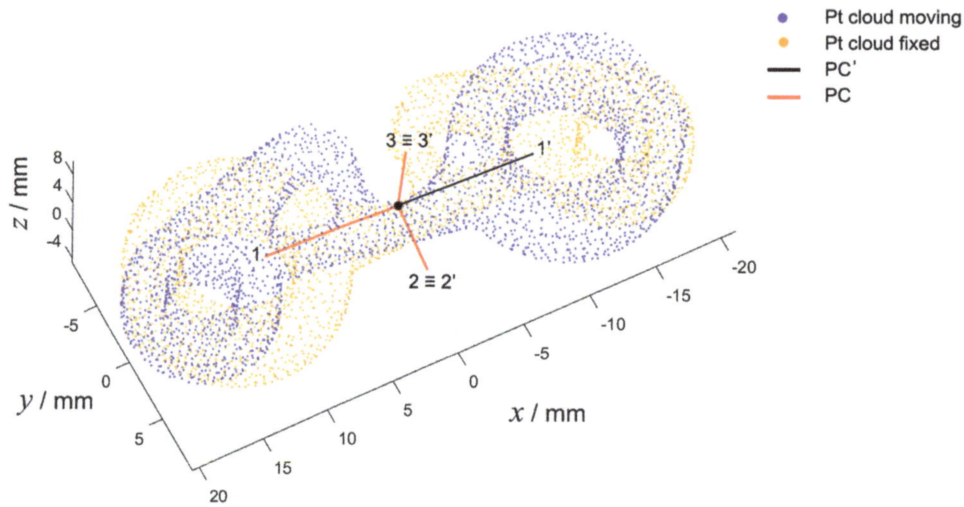

Figure 2.20. Example of wrong PCA alignment: the principal axis PC1' (black) of the moving point cloud (blue) is matched correctly with PC1 (red) of the fixed point cloud (orange). However, the directions are opposite (negative and positive, respectively).

The most popular approaches which implement the first solution (best pairwise matches are those leading to optimal alignment) are known as RANSAC, and iterative closest points (ICP) alignments. RANSAC alignment is discussed here, while ICP will be illustrated in section 2.6.5. Alignment methods based on similarity matches between landmarks are discussed in section 2.6.6.

The RANSAC alignment algorithm is conceptually very similar to RANSAC fitting, discussed in section 2.5.1.5—see also Fischler and Bolles (1981). The general idea in RANSAC alignment is to iteratively create random pairwise matches until a configuration is found that brings the majority of the points belonging to a moving point cloud **P** into alignment with a fixed point cloud **Q**. In detail, at each iteration, a set of random matches between **P** and **Q** is created involving only a core subset of the two point clouds. These matches are used to solve the absolute orientation problem, leading to the transformation of the entire **P** point cloud. Further pairwise matches are then constructed between the points not belonging to the core subset and evaluated for goodness of alignment (using the same equation (2.23) to solve the absolute orientation problem). If a sufficient number of pairs are found in good alignment (i.e. there is a sufficient consensus), then the iterations terminate, and the alignment is considered as valid. In contrast, if too few pairs show good alignment, the procedure is repeated, starting from a new random core subset. The advantage of RANSAC alignment is its simplicity of implementation and speed, as it relies on closed-form solution of the absolute orientation problem. The downside is that there is no guarantee that an optimal alignment is found, due to the inherent randomness of the method and due to the need to set the controlling parameters of the method in a trial and error fashion. An example of coarse alignment between two datasets, matching correspondences via RANSAC, is shown in figure 2.21.

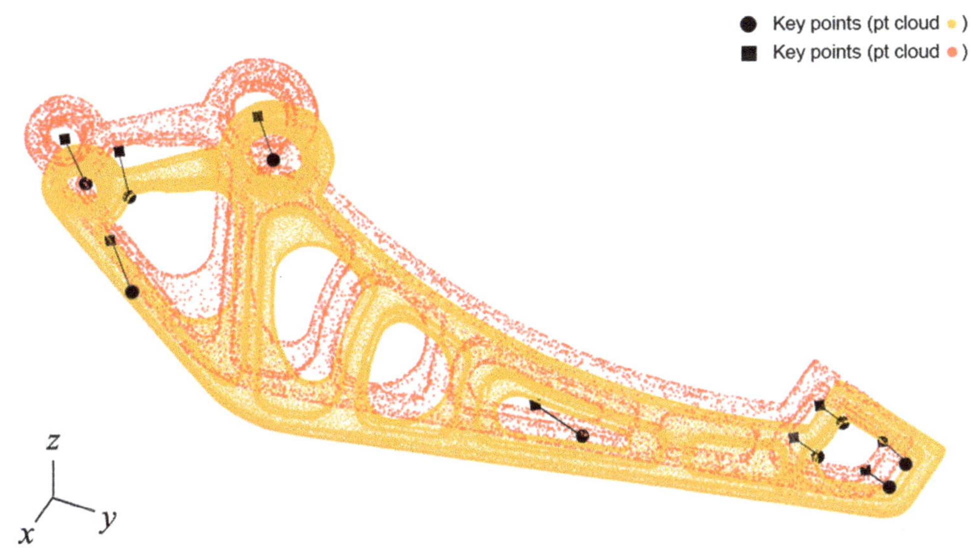

Figure 2.21. Paired of correspondences (squares and circles) identified in two point clouds (red and orange, respectively) are matched via RANSAC alignment.

2.6.5 Alignment by iterative closest points

The disadvantage of RANSAC alignment is that there is no learning—no evolution of the alignment result towards an increasingly better solution. In RANSAC alignment, if the consensus is not reached at any given iteration, the search restarts from the beginning, from an entirely new set of randomly selected, tentative correspondences. The iterative closest point (ICP) method (Besl and McKay 1992) circumvents this problem by introducing a different type of iteration, based instead on refinement of the previous result. There is actually a multitude of variants of ICP (Rusinkiewicz and Levoy 2001), so in the following section, only the most characteristic aspects are discussed.

As with RANSAC, the ICP method begins by creating a set of tentative matches, typically by selecting a random core subset of points in the moving **P** and fixed **Q** point clouds. The absolute orientation problem is solved, and **P** is brought into the best possible alignment for the given set of correspondences. Differently from RANSAC, after absolute orientation is solved, the correspondences in the core set are broken and an entirely new set is formed by searching for the nearest neighbour $\boldsymbol{q}$ of each point $\boldsymbol{p}_i$. When the new correspondences are established, the absolute orientation problem is solved again and the moving point cloud **P** is displaced once more. The process continues iteratively, with new correspondences being formed and a new absolute orientation problem being solved at each step. The termination criterion is usually related to the algorithm having reached a maximum number of iterations, or to the detection of the interruption of motion of the moving point cloud, meaning that final correspondences have been found and stabilised, so no further change is required.

Because of the intrinsic nature of the ICP iterative loop, the most significant displacement of the moving point cloud **P** occurs at the first iteration step, when the absolute orientation problem is solved the first time. After the first step, **P** can only shift by small amounts, corresponding to small changes in the set of correspondences and soon reaches a stop when the method is unable to evolve the correspondences any further. ICP is, therefore, very likely to quickly converge to a local minimum, with little capability of exploring the search space any further. ICP performance is strongly dependent on the initial position of **P** and thus is typically adopted as a refinement of a previous, coarse alignment step (typically performed by landmark matching, see section 2.6.6).

Several researchers have studied ways to increase the exploration power of ICP. For example, the use of an additional step to forcibly displace the moving point cloud by a small amount after resolution of the absolute orientation problem has been suggested to promote the exploration of new correspondences (Besl and McKay 1992). The creation of new, temporary points in the fixed cloud **Q** has also been proposed, again as a means to facilitate further motion of **P** after the first iteration step. Temporary points can be created by interpolation on **Q** (Besl and McKay 1992, Rusinkiewicz and Levoy 2001) or on tangent planes defined locally on the fixed surface (Chen and Medioni 1992, Bergevin *et al* 1996). If **Q** is comprised of vertices of a triangle mesh, temporary points can be created by interpolating directly on the triangle mesh (see figure 2.22).

In some circumstances, ICP may fail to converge—geometries that do not have enough constraints for good convergence are identified as unstable (Gelfand *et al* 2003). It follows that further challenges must be tackled to make ICP capable of detecting failure to obtain a stable result. Another challenge with ICP is related to processing speed, as the search for closest neighbours to establish correspondences at each step is slow and computationally expensive. In addition to point cloud partitioning methods dedicated to accelerating point search, such as *k*-NN (discussed in section 2.2.3.4), methods specifically dedicated to ICP have been developed, for example the projective association method (Blais and Levine 1995).

A recent overview of ICP variants is given elsewhere (Holz *et al* 2015). ICP variants, for example, introduce different ways to identify the initial random core

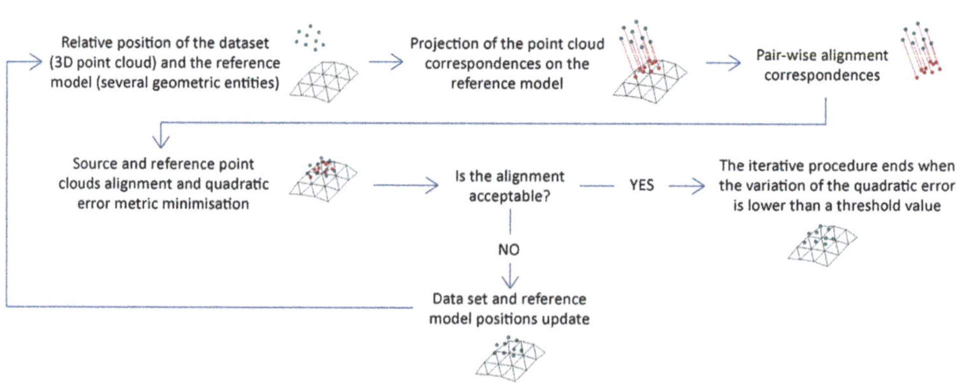

Figure 2.22. The iterative workflow of the ICP scheme. Registration between a point cloud and a triangle surface.

subset for the first alignment iteration, adopt different distance metrics to equation (2.23), present different methods to remove outlier points from the computation of distance at each iteration, or propose weighing methods to give different relative importance to specific correspondences.

2.6.6 Landmark matching and alignment using similarity metrics

As introduced in section 2.6.4, there are two fundamental concepts that can be adopted to identify correspondences and solve the alignment problem. The first concept is that the best matching points are those that can be better aligned, regardless of any other consideration about the nature of the selected points. Both RANSAC alignment and ICP alignment fundamentally follow this concept. The second concept is that the best matching points are those that are representative of similar regions, regardless of whether such matches can be optimally aligned or not. Alignment methods adopting the second concept require:

(a) The definition of criteria to locate 'interesting' regions in the point cloud; interesting because they could host candidate landmarks.
(b) The development of solutions to mathematically encode the shape-related information associated with such regions/landmarks.
(c) The development of solutions to quantitatively assess the degree of similarity (or difference) or paired regions/landmarks (Tombari *et al* 2010, Bellekens *et al* 2015, Han *et al* 2018).

When correspondences between landmarks have been established following the second concept, once again the alignment is typically achieved by resolution of the absolute orientation problem (for example, using the centroids of the paired landmark regions) (Szeliski 2011). There is a large variety of alignment methods that rely on the identification of matched landmarks. Any local shape descriptor, such as those described in section 2.4, for example local curvature, may be used to search for landmarks and later for pairwise matches based on similarity. Finally, it is worth pointing out that alignment methods may actually implement a hybrid combination of the two concepts. For example, in figure 2.23, candidate landmarks are identified by computing local surface curvatures. Because curvature similarity alone does not have sufficient discriminative power to allow robust identification of the strongest matches (different local surface shapes may have very similar curvature), the candidates have been processed using a RANSAC alignment procedure and refined with ICP (Catalucci *et al* 2019).

2.7 Measurement uncertainty in point cloud surface data

The term measurement uncertainty refers to a non-negative parameter characterising the dispersion of the quantity values being attributed to a measurand (BIPM 2012). In layman's terms, measurement uncertainty addresses our lack of knowledge of a real quantity value (distance, length, area, volume, etc) due to measurement error, and defines an interval, centred on the result of a measurement, within which the real quantity value is supposed to be located. The problem of correctly

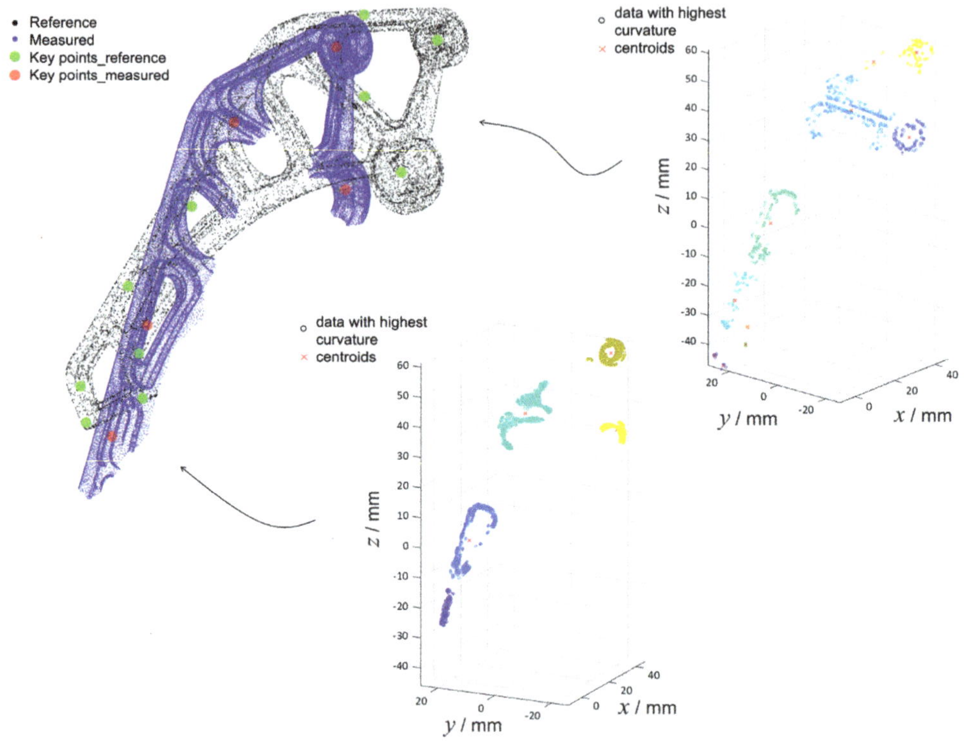

Figure 2.23. Data, divided into clusters, represent the points in the cloud with the highest Gaussian curvature values. The centroids of those subsets are matched via RANSAC.

quantifying measurement uncertainty for a given measurement process, or more generally, for any given measurement system, is still an open research question. For measurement involving the generation of point clouds, the challenge is to understand how measurement error influences the position of the points in the point cloud (positional uncertainty) (Forbes 2018) and how such uncertainty propagates through the data processing pipeline, ultimately affecting the results of the inspection or verification process. General methodological aspects related to how uncertainty contributions may be estimated and combined can be found in the *Guide to the Expression of Uncertainty in Measurement* (GUM) (JCGM 2008, Haitjema 2018, Leach and Shaheen 2020) for a discussion of measurement uncertainty for optical CMS.

2.7.1 Approaches to the estimation of measurement uncertainty

One approach to estimate measurement uncertainty consists of starting from first principles, i.e. starting from the detailed study of all possible error sources affecting a measurement pipeline, and the development of mathematical models representing each source (Barchiesi and Grosges 2017, Haitjema 2018). Error models are then aggregated through further mathematical modelling, with the ultimate goal of producing comprehensive analytical representations of uncertainty which may be applied to any circumstance, by applying the appropriate model parameters. The starting-from-first-principles approach has proven difficult for point clouds because of the multitude of possible error sources and the complexities of their interactions, leading into significant challenges for the mathematical modelling. Thus, existing models following this approach have been limited so far, for example providing models applicable to the measurement of elementary prismatic parts or simple smooth spheres, in most cases exclusively addressing measurement by contact-probing (i.e. tactile CMSs) (ISO 10360 series—see chapter 1 for details). Research on non-contact measurement is less mature, although recent work has begun highlighting the main issues for several optical technologies (Gayton *et al* 2019, Leach *et al* 2019, Lou *et al* 2019, Rivas Santos *et al* 2020, Leach and Shaheen 2020).

To overcome the complexity of uncertainty estimation from first principles, the ISO Technical Committee TC 213, Working Group 10 has proposed alternative methods. One of the main ideas is to focus on the quantification of traceability instead. Traceability refers to the comparison of a measurement result with a reference value, through a documented unbroken chain of passages each contributing with a known amount of measurement uncertainty (see the definition in chapter 1 and BIPM *et al* 2012).

Uncertainty quantification can in some cases be simplified by comparing the measurement to a similar measurement of a reference object. For example, a method has been proposed in ISO 15530 part 3 (ISO 2011), known as the 'substitution method'. In the substitution method, the uncertainty evaluation is based on a sequence of measurements on a calibrated object or objects, performed in the same way and under the same conditions as the actual measurements. The differences between the measurement of the object of interest and the measurement of the reference artefact are used as the starting point to estimate uncertainty.

A different approach, described in ISO 15530 part 4 (ISO/TS 2008), goes back to the idea of estimating uncertainty without the use of a reference object, but instead attempting to create mathematical models of the measurement process from first principles using simulation. Simulated models are constructed representing specific aspects of the measurement pipeline (either hardware or software components of the measurement system). Each part of the simulation is developed to address a limited set of parameters and describes a limited range of scenarios (i.e. 'task-specific' simulation), therefore accepting a limited range of values in its input parameters. The aggregation of simulation models leads to the creation of a 'virtual coordinate measuring system' (Balsamo *et al* 1999, Flack 2013, Leach and Shaheen 2020). Determination of measurement uncertainty using a virtual CMS is carried out by performing repeated simulated measurements with varying inputs (influence quantities) and determining how those inputs affect the measurand. Propagation through the modules comprised within the simulation is usually achieved through Monte Carlo methods (Haitjema 2018). The simulation is repeated a significant number of times until a statistical evaluation of these virtual measurements is made. Because typically the virtual CMS have been constructed to match specific scenarios, the results are often referred to as 'task-specific'.

Research on virtual CMS has so far concentrated on tactile CMS and there are commercial realisations. Initial work has been carried out to model virtual optical CMSs (Zhang *et al* 2018, Gayton *et al* 2019, Sims-Waterhouse *et al* 2020), but this is still an open research area, in particular given the challenges of developing reliable simulations to capture the interactions between light and the measured surface. Further details on uncertainty estimation based on the concept of virtual CMS can be found in Leach and Shaheen (2020).

2.7.2 Uncertainty associated with point clouds

Despite optical measurement technologies involving the generation and processing of point cloud surface data, few approaches devoted to estimating uncertainty in optical measurement are fully dedicated to understanding and modelling the uncertainty associated with the individual points of the point cloud. Such uncertainty, referred to as 'positional uncertainty', is the result of a chain of physical events taking place in the measurement process and its in-depth investigation would allow for better understanding of how measurement error propagates through the data processing pipeline influencing simplification, filtering, partitioning and datum fitting, and ultimately affecting the results of the characterisation process. The relationship between measurement uncertainty and fitting has also been investigated by Forbes (2006a, 2006b) and Forbes and Minh (2011) as a means to determine the uncertainty in the characterisation of form error.

Research on how to describe the spatial placement of points of a cloud in probabilistic terms has been carried out in relation to skin models in the field of geometric tolerancing (Anwer *et al* 2013, Zhang *et al* 2011, 2013). A skin model is a surface model capable of representing geometric variability across manufactured instances of the same part. For skin models, the idea of using random variables to describe the coordinates of each point has been explored. As shown in figure 2.24, in some cases, a random variable is only associated with a displacement in the direction defined by the local surface normal, in other cases, a full-3D probability ellipsoid (tri-variate random variable) is associated with each point.

Univariate random variables associated with local surface normals have been explored for example in Thompson *et al* (2017), specifically as a means of addressing measurement uncertainty. Random variables may be defined as independent between points or spatial dependency can be captured by modelling covariance (Zhang *et al* 2011). Typically, statistical point cloud models for addressing

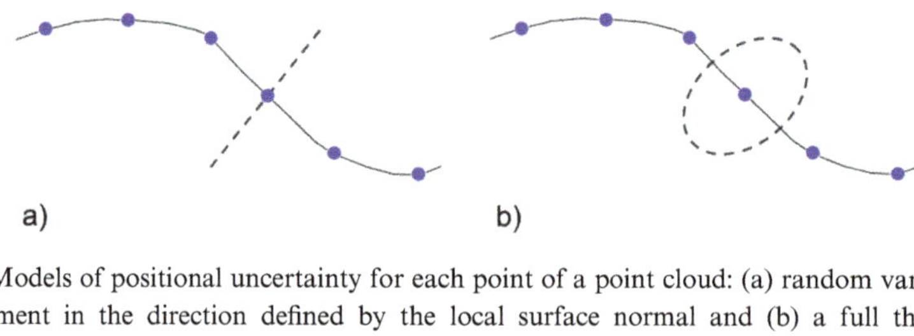

Figure 2.24. Models of positional uncertainty for each point of a point cloud: (a) random variable associated to a displacement in the direction defined by the local surface normal and (b) a full three-dimensional probability ellipsoid (tri-variate random variable) associated with each point.

measurement uncertainty are constructed by aggregating repeated measurements. For example, Senin *et al* (2020) present a method based on fitting Gaussian random fields to high-density point clouds produced by measurement repeats. The fitted field delivers a depiction of the spatial distribution of random measurement error over a part geometry and can incorporate local bias information through further measurement, or with the use of an external model to obtain a complete, spatial uncertainty map. Statistical models can also be obtained by fitting an individual measurement (single point cloud), as long as specific assumptions are made. For example, Evans (2008) has demonstrated how standard deviation maps computed from repeated height observations can be propagated into the assessment of peak-to-valley form error for a flat surface, using Monte Carlo simulation. Standard deviation maps obtained from measurement repeats are also useful for calibration and more general tasks related to uncertainty assessment (Evans 2010, Evans and Davies 2013). In Moroni *et al* (2018), a statistical model is obtained by fitting the point cloud to a Gaussian field using a shift-invariant model for covariance defined using a co-variogram. A variety of other approaches have been presented to represent point clouds in probabilistic terms, see for example, Pauly *et al* (2004), Chen *et al* (2016), Du *et al* (2016a, 2016b), Ozendi *et al* (2016), Niemeier and Tengen (2017).

The advantage of having a statistical model for a point cloud, is that many new 'virtual' observations of the cloud can be obtained by simulation, and error propagation through the data processing pipeline can be investigated without the need for analytical models. This is typically carried out using Monte Carlo simulation (Moroni *et al* 2018, Senin *et al* 2020) or Markov chain Monte Carlo simulation (Anwer *et al* 2013).

2.8 Conclusions

In optical metrology, point clouds play a central role—measurement systems produce point clouds, features of size are computed from point clouds and verification of conformance to specifications (for example, tolerances) are computed from point clouds.

In the last few decades, in particular due to non-contact measurement technologies, increasingly dense point clouds have begun to be available in coordinate measurement, providing the possibility of investigating the geometrical properties of manufactured parts at an unprecedented range of scales, thus providing a competitive advantage in the inspection of complex geometries, such as those obtainable via additive manufacturing. In contemporary coordinate measurement systems, measurement technologies and point cloud data processing technologies play an equally important role; this chapter has focused on the latter. The computational pipeline that starts from the point cloud and leads to the result of an inspection or verification task is complex and often comprised of numerous steps. Point cloud simplification, resampling, elimination of isolated points, noise reduction, segmentation and partitioning, fitting, registration and more, are all necessary steps that have been discussed. The mathematical methods and algorithms underlying the aforementioned steps are numerous and the subject of active research. Many solutions are rapidly evolving to cope with the increase in density of the point clouds and to take advantage of the rapidly increasing computational capabilities offered by current systems and technologies. Despite the recent advances, the role of data processing, which influences the final inspection result while in turn being influenced by error in the point cloud, has not yet been comprehensively investigated.

The recognition of the equally important role played by measurement technologies and point cloud data processing technologies in industrial inspection is the starting point to reach a more comprehensive understanding of the behaviour and performance of coordinate measurement for inspection and verification. In such a context, probably the most difficult challenge that must be tackled by research in point cloud data analysis and processing is related to increasing our understanding of the roles and effects of measurement error. Error is embedded in the point cloud resulting from measurement; error propagates through simplification, partitioning, fitting, registration, ultimately influencing the final result. A full understanding of such effects, combined with the identification of suitable ways to express the findings in terms of measurement uncertainty, is key to the advancement of coordinate metrology.

References

Abdi H 2009 Centroids *Wiley Interdiscip. Rev. Comput. Stat.* **1** 259–60

Aldoma A, Marton Z C, Tombari F, Wohlkinger W, Potthast C, Zeisl B, Rusu R, Gedikli S and Vincze M 2012 Tutorial: point cloud library: three-dimensional object recognition and 6 DOF pose estimation *Robot. Autom. Mag.* **19** 80–91

Alexa M, Gross M, Pauly M, Pfister H, Stamminger M and Zwicker M 2004 Point-based computer graphics *ACM SIGGRAPH '04 Course Notes (Los Angeles, CA, USA)*

Aluru S 2004 Quadtrees and octrees *Handbook of Data Structures and Applications* 2nd edn ed D P Mehta and S Sahni (London: Chapman and Hall/CRC)

Anwer N, Ballu A and Mathieu L 2013 The skin model, a comprehensive geometric model for engineering design *Ann. CIRP* **62** 143–6

Arun K S, Huang T S and Blostein S D 1987 Least-squares fitting of two 3-D point sets *IEEE Trans. Pattern Anal. Mach. Intell.* **9** 698–700

Balsamo A, Di Ciommo M, Mugno R, Rebaglia B I, Ricci E and Grella R 1999 Evaluation of CMM uncertainty through Monte Carlo simulations *Ann. CIRP* **48** 425–8

Bär C 2010 *Elementary Differential Geometry* (Cambridge: Cambridge University Press)

Barchiesi D and Grosges T 2017 Propagation of uncertainties and applications in numerical modeling: tutorial *J. Opt. Soc. Am.* A **34** 1602–19

Bellekens B, Spruyt V, Berkvens R and Penne R 2015 A benchmark survey of rigid 3D point cloud registration algorithms *IARIA* **8** 118–27

Bentley J L 1975 Multidimensional binary search trees used for associative searching *Commun. ACM* **18** 509–17

Berger M, Tagliasacchi A, Seversky L M, Alliez P, Levine J A, Sharf A and Silva C T 2014 State of the art in surface reconstruction from point clouds *Proc. EG '14 (Strasbourg, France)* pp 161–85

Berger M, Tagliasacchi A, Seversky L M, Alliez P, Guennebaud G, Levine J A, Sharf A and Silva C T 2017 A survey of surface reconstruction from point clouds *Comput. Graph. Forum* **36** 301–29

Bergevin R, Soucy M, Qagnon H and Laurendeau D 1996 Towards a general multi-view registration technique *IEEE Trans. Pattern Anal. Mach. Intell.* **18** 540–7

Besl P and McKay N 1992 A method for registration of 3-D shapes *IEEE Trans. Pattern Anal. Mach. Intell.* **14** 239–56

BIPM, IEC, IFCC, ILAC, ISO, IUPAC, IUPAP and OIML 2012 *International Vocabulary of Metrology—Basic and General Concepts and Associated Terms—JCGM 200* (Geneva: Bureau International des Poids et Mesures)

Blais G and Levine M D 1995 Registering multiview range data to create 3D computer objects *IEEE Trans. Pattern Anal. Mach. Intell.* **17** 820–4

Bommes D, Lévy B, Pietroni N, Puppo E, Silva C and Tarini M 2013 Quad-mesh generation and processing: a survey *Comput. Graph. Forum* **32** 51–76

Brown B J and Rusinkiewicz S 2007 Global non-rigid alignment of 3-D scans *ACM Trans. Graph.* **26** 21

Catalucci S, Senin N, Piano S and Leach R K 2019 Intelligent systems for optical form measurement: automated assessment of pose and coverage *Proc. 34th ASPE (Pittsburgh, PA, USA)* pp 145–50

Chen X, Zhang G, Zhang J, Wu H, Lu T and Xuan W 2016 An average error ellipsoid model for evaluating precision of point cloud from TLS *J. Ind. Soc. Remote Sens* **44** 865–73

Chen Y and Medioni G 1992 Object modelling by registration of multiple range images *Image Vis. Comput.* **10** 145–55

De Berg M, Cheong O, Van Kreveld M and OverMars M 2008 *Computational Geometry: Algorithms and Applications* 3rd edn (Berlin: Springer)

Du Z, Wu Z and Yang J 2016a Error ellipsoid analysis for the diameter measurement of cylindroid components using a laser radar measurement system *Sensors* **16** 714

Du Z, Wu Z and Yang J 2016b Point cloud uncertainty analysis for laser radar measurement system based on error ellipsoid model *Opt. Laser Eng.* **79** 78–84

Dyn N, Hormann K, Kim S J and Levin D 2001 Optimizing 3D triangulations using discrete curvature analysis *Mathematical Methods for Curves and Surfaces* ed T Lyche and L L Schumaker (Nashville,TN: Vanderbilt University)

Edelsbrunner H 2009 Delaunay triangulations *Geometry and Topology for Mesh Generation* ed H Edelsbrunner (Cambridge: Cambridge University Press)

Eggert D W, Lorusso A and Fisher R B 1997 Estimating 3-D rigid body transformations: a comparison of four major algorithms *Mach. Vision Appl.* **9** 272–90

Elseberg J, Magnenat S, Siegwart R and Andreas N 2012 Comparison of nearest-neighbor-search strategies and implementations for efficient shape registration *J. Softw. Eng. Robot.* **3** 2–12

Evans C J 2008 Uncertainty evaluation for measurements of peak to valley surface form errors *Ann. CIRP* **57** 509–12

Evans C J 2010 Certification, self-calibration, and uncertainty in testing optical flats *Proc. SPIE* **7656** 76560S

Evans C and Davies A 2013 Certification, self-calibration and uncertainty in optical surface testing *Int. J. Precis. Technol.* **3** 388–401

Feng S C and Hopp T 1991 A review of current geometric tolerancing theories and inspection data analysis algorithms *NIST Interagency/Internal Report* (NISTIR-4509)

Fischler M A and Bolles R C 1981 Random sample consensus: a paradigm for model fitting with applications to image analysis and automated cartography *Commun. ACM* **24** 381–95

Fisher R B 1989 Description of three dimensional structures *From Surface To Objects: Computer Vision and Three Dimensional Scene Analysis* ed R B Fisher (Chichester: Wiley)

Flack D 2013 *Measurement Good Practice Guide No. 130: Co-ordinate Measuring Machine Task-Specific Measurement Uncertainties* (London: National Physical Laboratory)

Forbes A B 2006a Surface fitting taking into account uncertainty structure in coordinate data *Meas. Sci. Technol.* **17** 553–8

Forbes A B 2006b Uncertainty evaluation associated with fitting geometric surfaces to coordinate data *Metrologia* **43** 282–90

Forbes A B 2018 Uncertainties associated with position, size and shape for point cloud data *J. Phys. Conf. Ser.* **1065** 142023

Forbes A B and Minh H D 2011 Form assessment in coordinate metrology *Approximation Algorithms for Complex Systems* (Springer Proceedings in Mathematics vol 3) ed E Georoulis, A Iske and J Levesley (Berlin: Springer) pp 69–90

Forbes A B and Minh H D 2012 Generation of numerical artefacts for geometric form and tolerance assessment *Int. J. Metrol. Qual. Eng.* **3** 145–50

Gayton G, Su R, Leach R K and Bradley-Smith L 2019 Uncertainty evaluation of fringe projection based on linear systems theory *Proc. 35th CMSC (Orlando, FL, USA)*

Gelfand N, Ikemoto L, Rusinkiewicz S and Levoy M 2003 Geometrically stable sampling for the ICP algorithm *Proc. 3DIM (Banff, AB, Canada)* pp 260–7

Grilli E, Menna F and Remondino F 2017 A review of point clouds segmentation and classification algorithms *ISPRS Arch.* **XLII-2/W3** 339–44

Gruen A and Akca D 2005 Least squares 3D surface and curve matching ISPRS *J. Photogramm. Remote Sens* **59** 151–74

Haitjema H 2018 Measurement uncertainty *Basics of Precision Engineering* ed R K Leach and S T Smith (Boca Raton, FL: CRC Press)

Han X F, Jin J S, Wang M J, Jiang W, Gao L and Xiao L 2017 A review of algorithms for filtering the 3D point cloud *Signal Process. Image Commun.* **57** 103–12

Han X F, Jin J S, Xie J, Wang M J and Jiang W 2018 A comprehensive review of 3D point cloud descriptors (arXiv: 1802.02297)

Holz D, Ichim A E, Tombari F, Rusu R B and Behnke S 2015 Registration with the point cloud library: a modular framework for aligning in 3-D *Robot. Autom. Mag.* **22** 110–24

Holzer S, Rusu R B, Dixon M, Gedikli S and Navab N 2012 Adaptive neighborhood selection for real-time surface normal estimation from organized point cloud data using integral images *IEEE/RSJ Int. Conf. on Intelligent Robots and Systems, (Vilamoura)* pp 2684–9

Hoppe H, DeRose T, Duchamp T, McDonald J and Stuetzle W 1992 Surface reconstruction from unorganized points *Comput. Graph. (ACM)* **26** 71–8

Horn B K P 1987 Closed-form solution of absolute orientation using unit quaternions *J. Opt. Soc. Am.* A **4** 629–42

Horn B K P, Hilden H M and Negahdaripour S 1988 Closed-form solution of absolute orientation using orthonormal matrices *J. Opt. Soc. Am.* A **5** 1127–35

Hosoi F and Omasa K 2006 Voxel-based 3-D modeling of individual trees for estimating leaf area density using high-resolution portable scanning lidar *IEEE Trans. Geosci. Remote Sens* **44** 3610–8

Huber D 2011 The ASTM E57 file format for 3D imaging data exchange *Proc. SPIE* **7864** 78640A

ISO 15530 part 3 2011 *Geometrical Product Specifications (GPS)—Coordinate Measuring Machines (CMM): Technique for Determining the Uncertainty of Measurement—Part 3: Use of calibrated workpieces or standards* (Geneva: International Organization for Standardization)

ISO/TS 15530 part 4 2008 *Geometrical Product Specifications (GPS)—Coordinate Measuring Machines (CMM): Technique for Determining the Uncertainty of Measurement—Part 4: Evaluating task-specific measurement uncertainty using simulation* (Geneva: International Organization for Standardization)

James G, Witten D, Hastie T and Tibshirani R 2013 *An Introduction to Statistical Learning with Applications in R* (New York: Springer)

JCGM 2008 *GUM 1995 with Minor Corrections, Evaluation of Measurement Data—Guide to the Expression of Uncertainty in Measurement—JCGM 100* (Sèvres: BIPM)

Jiménez P, Thomas F and Torras C 2001 3D collision detection: a survey *Comput. Graph.* **25** 269–85

Jolliffe I T 2002 *Principal Component Analysis* (New York: Springer)

Kendall D G 1989 A survey of the statistical theory of shape *Stat. Sci.* **4** 87–99

Kobbelt L and Botsch M 2004 A survey of point-based techniques in computer graphics *Comput. Graph.* **28** 801–14

Landa J, Prochazka D and Štastny J 2013 Point cloud processing for smart systems *Acta Univ. Agric. Silvic. Mendel. Brun* **61** 2415–21

Leach R K, Bourell D, Carmignato S, Donmez A, Senin N and Dewulf W 2019 Geometrical metrology for metal additive manufacturing *Ann. CIRP* **68** 677–700

Leach R K and Shaheen A 2020 Post-process coordinate metrology *Precision Metal Additive Manufacturing* ed R K Leach and S Carmignato (Boca Raton, FL: CRC Press)

Linsen L 2001 Point cloud representation *University of Karlsruhe, Germany Technical Report, Faculty of Informatics*

Lou S, Brown S B, Sun W, Zeng W, Jiang X and Scott P J 2019 An investigation of the mechanical filtering effect of tactile CMM in the measurement of additively manufactured parts *Measurement* **144** 173–82

Lourakis M 2016 An efficient solution to absolute orientation *Proc. 23rd Int. Conf. Pattern Recogn. (Cancun, Mexico)* pp 3816–9

Magid E, Soldea O and Rivlin E 2007 A comparison of Gaussian and mean curvature estimation methods on triangular meshes of range image data *Comput. Vis. Image Underst* **107** 139–59

Mérigot Q, Ovsjanikov M and Guibas L J 2011 Voronoi-based curvature and feature estimation from point clouds *IEEE Trans. Vis. Comput. Graph.* **17** 743–56

Meyer M, Desbrun M, Schröder P and Barr A H 2003 Discrete differential-geometry operators for triangulated 2-manifolds *Visualization and Mathematics III* ed H C Hege and K Polthier (Berlin: Springer)

Monserrat O and Crosetto M 2008 Deformation measurement using terrestrial laser scanning data and least squares 3D surface matching ISPRS *J. Photogramm. Remote Sens.* **63** 142–54

Moroni G and Petrò S 2008 Geometric tolerance evaluation: a discussion on minimum zone fitting algorithms *Precis. Eng.* **32** 232–7

Moroni G, Syam W P and Petrò S 2018 A simulation method to estimate task-specific uncertainty in 3D microscopy *Measurement* **122** 402–16

Nguyen A and Le B 2013 3D point cloud segmentation: a survey *6th IEEE RAM Proc. (Manila, Philippines)* pp 225–30

Niemeier W and Tengen D 2017 Uncertainty assessment in geodetic network adjustment by combining GUM and Monte-Carlo simulations *J. Geod.* **11** 67–76

Owen S 1998 A survey of unstructured mesh generation technology *7th Int. Meshing Roundtable* pp 239–67

Ozendi M, Akca D and Topan H 2016 An emprical point error model for TLS derived point clouds *ISPRS Arch.* **XLI-B5** 557–63

Pauling F, Bosse M and Zlot R 2009 Automatic segmentation of 3D laser point clouds by ellipsoidal region growing *Proc. ACRA '09 (Sydney, Australia)*

Pauly M, Gross M and Kobbelt L P 2002 Efficient simplification of point-sampled surfaces *Proc. IEEE VIS '02 (Boston, MA, USA)* pp 163–70

Pauly M, Mitra N J and Guibas L 2004 Uncertainty and variability in point cloud surface data *SPBG '04 (Goslar, Germany)* pp 77–84

Pressley A 2010 Curvature of surfaces *Elementary Differential Geometry* ed A Pressley (London: Springer)

Pulli K 1999 Multiview registration for large data sets *Proc. 3DIM (Ottawa, ON, Canada)* pp 160–8

Rabbani T, van den Wildenberg F and Vosselman G 2006 Segmentation of point clouds using smoothness constraint *Int. Arch. Photogramm. Remote Sens. Spat. Inf. Sci.* **35** 248–53

Ramasamy S K, Raja J and Boudreau B D 2013 Data fusion strategy for multiscale surface measurements *J. Micro Nano-Manuf.* **1** 011004

Rivas Santos V M, Thompson A, Sims-Waterhouse D, Maskery I, Woolliams P and Leach R K 2020 Design and characterisation of an additive manufacturing benchmarking artefact following a design-for-metrology approach *Addit. Manuf.* **32** 100964

Rusinkiewicz S and Levoy M 2001 Efficient variants of the ICP algorithm *Proc. 3DIM (Quebec, QC, Canada)* pp 145–52

Rusu R B 2009 Semantic 3D object maps for everyday manipulation in human living environments *PhD Thesis* Technische Universität Müchen, Germany

Rusu R B and Cousins S 2011 3D is here: point cloud library *IEEE ICRA* pp 1–4

Salvi J, Matabosch C, Fofi D and Forest J 2007 A review of recent range image registration methods with accuracy evaluation *Image Vision Comput.* **25** 578–96

Samet H 1989 *Applications of Spatial Data Structures: Computer Graphics, Image Processing, and GIS* (Boston, MA: Addison-Wesley/Longman)

Samet H 1990 *The Design and Analysis of Spatial Data Structures* (Boston, MA: Addison-Wesley/Longman)

Saval-Calvo M, Orts-Escolano S, Azorin-Lopez J, García Rodríguez J, Fuster-Guillo A, Morell-Gimenez V and Cazorla M 2015 A comparative study of downsampling techniques for non-rigid point set registration using color *IWINAC* pp 281–90

Schall O, Belyaev A and Seidel H P 2005 Robust filtering of noisy scattered point data *Symp. Point-Based Graphics '05 (Stony Brook, NY, USA)* pp 71–7

Schall O, Belyaev A and Seidel H P 2008 Adaptive feature-preserving non-local denoising of static and time-varying range data *Comput. Aided Des.* **40** 701–7

Schnabel R, Wahl R and Klein R 2007 Efficient RANSAC for point-cloud shape detection *Comput. Graph. Forum* **26** 214–26

Schneider P J and Eberly D H 2003 *Geometric Tools for Computer Graphics* (San Francisco, CA: Morgan Kaufmann)

Senin N, Catalucci S, Moretti M and Leach R K 2020 Statistical point cloud model to investigate measurement uncertainty in coordinate metrology *Accepted*

Shapiro V 2002 Solid modeling *Handbook of Computer Aided Geometric Design* ed G Farin, J Hoschek and M S Kim (Amsterdam: Elsevier) pp 473–518

Sims-Waterhouse D, Isa M, Piano S and Leach R K 2020 Uncertainty model for a traceable stereo-photogrammetry system *Precis. Eng.* **63** 1–9

Sotoodeh S 2006 Outlier detection in laser scanner point clouds *ISPRS Arch.* **XXXVI-5** 297–302
Strutz T 2016 *Data Fitting and Uncertainty. A Practical Introduction to Weighted Least Squares and Beyond* (Wiesbaden: Vieweg+Teubner)
Sultani Z N and Ghani R F 2015 Kinect 3D point cloud live video streaming *Procedia Comput. Sci.* **65** 125–32
Sun J, Jiang Y, Jiang J and Bai X 2016 Triangular mesh construction based on point cloud matrix and edge feature extraction *IEEE ITNEC* pp 275–9
Szeliski R 2011 *Computer Vision: Algorithms and Applications* (London: Springer)
Tam G K L, Cheng Z Q, Lai Y K, Langbein F C, Liu Y, Marshall D, Martin R R, Sun X F and Rosin P L 2013 Registration of 3D point clouds and meshes: a survey from rigid to nonrigid *IEEE T. Vis. Comput. Gr* **19** 1199–217
Taubin G 1995 Signal processing approach to fair surface design *ACM SIGGRAPH '95 (New York, NY, USA)* pp 351–8
Thompson A, Senin N, Giusca C and Leach R K 2017 Topography of selectively laser melted surfaces: a comparison of different measurement methods *Ann. CIRP* **66** 543–6
Tombari F, Salti S and Di Stefano L 2010 Unique signatures of histograms for local surface description *Computer Vision—ECCV 2010* (Lecture Notes in Computer Science vol 6313) ed K Daniilidis, P Maragos and N Paragios (Berlin: Springer)
Turk G 1994 *The PLY Polygon File Format* (Berkeley, CA: Stanford University)
van den Bergen G 1997 Efficient collision detection of complex deformable models using AABB trees *J. Graph. Tools* **2** 1–13
Wang J, Leach R K and Jiang X 2015 Review of the mathematical foundations of data fusion techniques in surface metrology *Surf. Topogr. Metrol. Prop.* **3** 023001
Weckenmann A, Jiang X, Sommer K D, Neuschaefer-Rube U, Seewig J, Shaw L and Estler T 2009 Multisensor data fusion in dimensional metrology *Ann. CIRP* **58** 701–21
Weinmann M 2016 Preliminaries of 3D point cloud processing *Reconstruction and Analysis of 3D Scenes* ed M Weinmann (Berlin: Springer)
Xiaohui D, Baocai Y and Dehui K 2007 Adaptive out-of-core simplification of large point clouds *Proc. IEEE ICME (Beijing, China)* pp 1439–42
Xie Y, Tian J and Zhu X X 2020 Linking points with labels in 3D: a review of point cloud semantic segmentation *IEEE Geosci. Rem. Sen. Mag.* (arXiv: 1908.08854)
Zhang B, Davies A, Evans C and Ziegert J 2018 Validity of the instrument transfer function for fringe projection metrology *Appl. Opt.* **57** 2795–803
Zhang M, Anwer N, Mathieu L and Zhao H B 2011 A discrete geometry framework for geometrical product specifications *Proc. CIRP Design Conf. (Korea)*
Zhang M, Anwer N, Stockinger A, Mathieu L and Wartzack S 2013 Discrete shape modeling for skin model representation *Proc. Inst. Mech. Eng.* B **227** 672–80

IOP Publishing

Chapter 3

Laser triangulation

Mohammed A Isa, Samanta Piano and Richard Leach

As one of the popular techniques for non-contact coordinate metrology, laser triangulation has developed from a method of range measurement to three-dimensional coordinate measurement. The growth in the application of laser triangulation sensors is credited to their adaptability to measurement requirements and versatility for distance and coordinate measurements. In addition, the robust and compact nature of laser triangulation sensors enables their integration into existing systems. Therefore, the laser triangulation principle has been applied for measuring sensors found in consumer products, industrial devices, medical equipment and transportation systems. This chapter gives a summary of laser triangulation as a measurement technique and outlines recent applications and challenges.

3.1 Laser triangulation

Laser triangulation (LT) is an active non-contact measurement strategy used by several optical distance and coordinate measurement systems (CMSs). LT systems carry out one-dimensional (1D) to three-dimensional (3D) coordinate measurements depending on the features that need to be measured. Object sizes ranging from small objects with microscale surface topography (Mueller *et al* 2015) to large-scale aircraft parts (Zhang *et al* 2015) can be measured using LT. LT is 'active' because of the requirement for an illumination source(s) in addition to the existing ambient illumination. Essentially, the principle of LT relies on the projection of structured laser illumination onto a surface area where the reflected light—carrying some topographical information about the incident area—is observed. A photodetector is used to capture the intensity of the reflected laser illumination received from the incident region. The distribution of the intensity values recorded by the photodetector is a result of the interaction between the laser illumination and the surface. The nature of this interaction depends on the topography of the surface and the properties of the propagating light.

doi:10.1088/978-0-7503-2524-0ch3

The origin of LT follows the nascent developments made in laser and photodetector technologies in the 1960s and 1970s. Propelled by the demand for non-invasive inspection methods, the period witnessed the exploration of optical methods consisting of scanning microscopy, interferometric techniques, fringe projection and laser scanning (Costa 2012). These discoveries were supported by advances in computing technologies and new electronic frameworks. Among the earliest research in non-invasive methods, triangulation-based distance and profile measuring systems were developed using laser light for non-contact measurement (Sawatari 1976, Smolka and Caudell 1978). The limits on the resolution of LT measurements were studied (Lim and Nawab 1981) and optical configurations using synchronised scanning approaches were introduced to increase the resolution of measurement (Rioux 1984). For improvement of the light detection in LT systems, enhanced analogue sensors were introduced before the advent of sophisticated digital sensors (Bertani *et al* 1984). Despite early improvements, the industrial use of LT for dimensional metrology was limited because contact-based and interferometric methods were generally preferred. LT was not considered to be a major measurement method and specification standards in dimensional metrology were drafted based on other measurement systems—stylus contact systems and interferometric systems (Costa 2012). However, compact and adaptable hardware made LT an attractive non-contact measurement method (Petrov *et al* 1998).

In the twenty-first century, the accuracy of LT was improved and its application in dimensional measurement was enhanced. The desire for portability and compactness, coupled with advances in digital imaging, laser technology and fast computing, steered LT into the industrial environment. LT became more frequently employed for in-process and coordinate metrology, particularly in measurements of car bodies in the automotive industry (Schwenke *et al* 2002). By 2010, over 36% of 3D measurement systems used in quality control departments were based on LT techniques (Reiner and Stankiewicz 2011). Even though LT might not be suitable for the inspection of the surface texture of high-precision manufacturing processes (Black and Kohser 2011), the specified tolerances at many stages of production are of the order of tens to hundreds of micrometres, which can make LT a suitable measurement method. To date, LT is the most commonly applied non-contact sensor in dimensional metrology and quality inspection (Brosed *et al* 2011, Cajal *et al* 2015, Martínez *et al* 2010), and remains a major technique that continues to be studied and improved (Schwarte *et al* 1999, Du and Xi 2019).

This chapter begins by introducing the principle of LT through analysis of a canonical range sensor. The influence of surface properties on LT measurements is discussed in section 3.3, along with the effects of speckle formation, surface reflectance and surface form on the measurement. Extensions of the LT principle to contemporary 2D and 3D CMSs are discussed in section 3.4. To understand the complete workings of modern LT systems, section 3.5 describes the general procedures involved in point cloud reconstruction from images. Finally, section 3.6 covers geometric inspection from cloud points and elaborates on the applications of reconstructed computer models.

3.2 Laser triangulation sensors

A typical LT sensor is composed of a collimated light source, light guiding optics and a photosensitive detector. These components can be packed into a portable LT sensor unit. The transmitted laser light interacts with the incident surface and a portion of the resulting scattered light is observed by the photodetector.

The most common light source for LT is a laser, which is usually a continuous or pulsed semiconductor laser diode. Typical lasers have radiant powers ranging from 1 to 100 mW, and the laser power is chosen based on the visibility at the desired working distance. Laser light at long working ranges, compared to close ranges, requires higher power lasers to be visible. Depending on the application, the incident laser beam is commonly structured to a specific shape, such as a spot or a sheet of light. The diameter of the laser spot or the thickness of the sheet of light are relatively small (usually between 10 and 500 μm), compared to the size of the measured objects, to allow sampling of narrow sections of surfaces. The spectrum of the light is chosen to be different from the general ambient light present in the environment of application. The low beam divergence and the small spectral bandwidth of lasers make them suitable for triangulation measurements (Donges and Noll 2015). In addition, coherent light from a laser can be collimated to keep the shape of illumination uniform across a measurement range.

At the light receiving end of an LT sensor, a photodetector is used to collect the light observed through the focusing lens. Analogue position sensitive detectors (PSDs), such as lateral-effect photodiodes (LEPs) or digital semiconductor sensors (CCDs and CMOSs) can be used as the photodetector. The combination of the lens and the digital photodetector forms a camera capable of storing light intensity data in one- (or two-) dimensional arrays of pixels. The use of digital sensors over analogue PSDs is due to the higher accuracy that can be obtained with them image processing (Schöch and Savio 2019).

The term 'triangulation' alludes to the geometric triangle formed by the incident laser beam, the observed reflected beam and the baseline distance along x from the projection centre point C to the laser beam, as shown in figure 3.1. This arrangement of optical triangulation is not unique to laser and structured light triangulation sensors, as it is also employed by passive stereoscopic and photogrammetric systems. Passive triangulation is discussed in chapter 4 of this book.

Referring to figure 3.1, the lateral distance x and depth z of the projected laser light on the surface can be evaluated using basic triangulation principles. The peak of the detected laser position u corresponds to the peak of the reflected laser beam. It is necessary to relate the detector position acquired from the LT sensor with the metric distance of the laser spot from the projection point C. Considering the central projection of the light rays at the focusing point C, the lateral position of the laser spot x is related to the depth z by

$$x = \frac{u}{f_x} z, \tag{3.1}$$

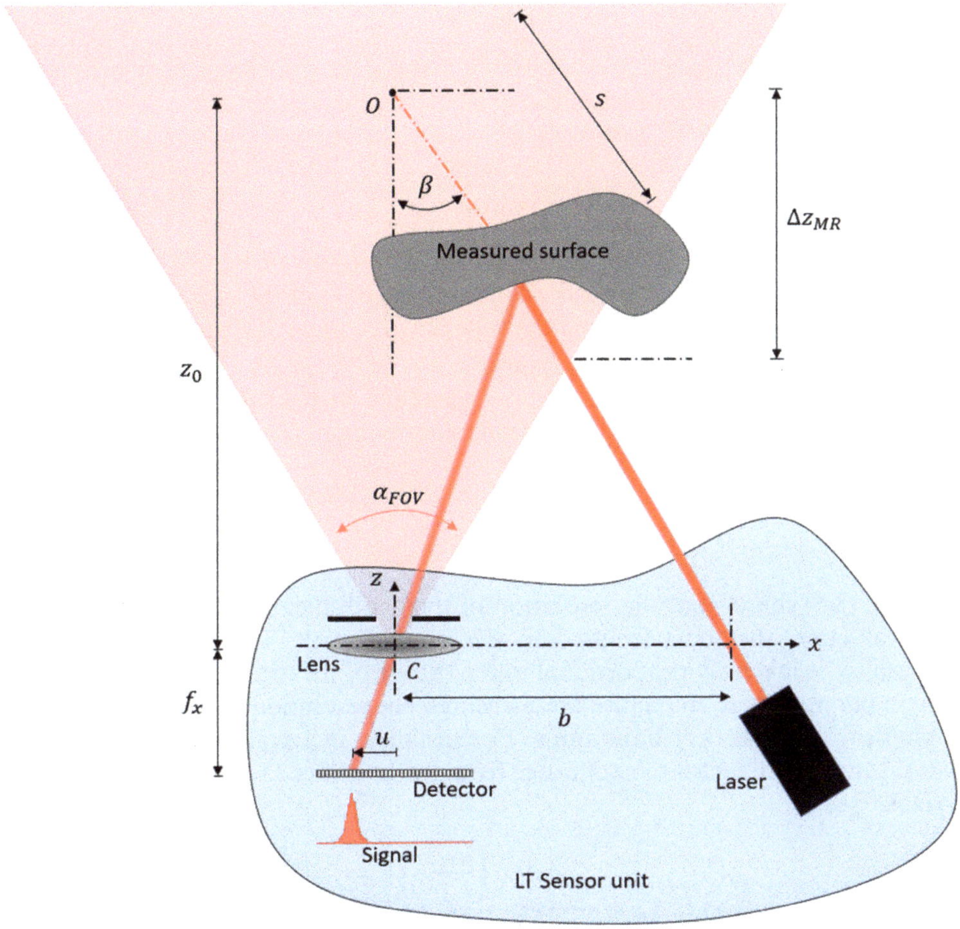

Figure 3.1. A canonical LT sensor unit showing light emitting and receiving components for coordinate measurement. Terms are explained in the main text.

where f_x is the focal length scaled to the detector pixel unit. The focal length in pixels is expressed as $f_x = fs_x$, where f is the metric focal length and s_x is the number of pixels per unit length (pixel density) of the detector. A geometrical relation from figure 3.1 can be obtained by triangulation as

$$x = (z_0 - z)\tan(\beta). \tag{3.2}$$

The triangulation angle β is the angle between the observation axis and the laser beam, where their intersection is represented by the point O in figure 3.1. The value z_0 is the reference depth which is the distance between points C and O along the observation axis. By rearranging equations (3.1) and (3.2) for the values of z and x, the depth can be expressed as

$$z = \frac{f_x \tan(\beta)}{f_x \tan(\beta) + u} z_0, \tag{3.3}$$

and the lateral position by

$$x = \frac{u \tan(\beta)}{f_x \tan(\beta) + u} z_0. \tag{3.4}$$

Other researchers evaluate the values of z and x in terms of the baseline distance b and laser angle $\beta' = \frac{\pi}{2} - \beta$ instead of the reference depth z_0 and the triangulation angle β (Peiravi and Taabbodi 2010, Idrobo-Pizo *et al* 2019). The baseline and the reference depth are related by $b = z_0 \tan(\beta)$.

The rate of change of detected pixel position with respect to the change in measured depth is defined as the triangulation gain (Pears 1994) and can be evaluated from equation (3.3) as

$$\left| \frac{\partial u}{\partial z} \right| = \frac{z_0 \tan(\beta)}{z^2} f_x = \frac{b}{z^2} f_x. \tag{3.5}$$

Equation (3.5) shows that the resolution of the depth measurement is proportional to the baseline, the focal length and the pixel density. The depth resolution is observed to be inversely proportional to the square of the measured depth. Selection of these parameters to maximise the measurement resolution is usually constrained by physical and practical limitations. For instance, increasing the baseline diminishes the maximum measurement range from the reference O in figure 3.1, according to (Isa 2018)

$$\Delta z_{\mathrm{MR}} = \frac{\tan\left(\frac{\alpha_{\mathrm{FOV}}}{2}\right)}{\tan\left(\frac{\alpha_{\mathrm{FOV}}}{2}\right) + \tan(\beta)} z_0. \tag{3.6}$$

Hence, the choice of baseline is limited by the required measurement range, which depends on the depth of focus, and the maximum range given by equation (3.6), where α_{FOV} is the lateral field of view. While pixel density is limited by manufacturing capacity and cost, the choice of focal length and measurement depth depends on the specific application requirements.

There are many variations to the canonical LT sensor discussed in this section; some of the common variants are covered in section 3.4, which also covers various scanning mechanisms that allow positioning of the LT sensor during a measurement process.

3.3 Laser triangulation measurement dependence on surface properties

This section covers the theoretical analysis of laser and surface factors—such as speckle formation, inclination angle and surface reflectance—that affect LT.

The optical property of the measured surface is known to be the most important contributor of uncertainty in LT (Schwenke *et al* 2002).

3.3.1 Measurement uncertainty limit

One of the limiting factors of optical systems is resolution, which determines the minimum lateral distance between features that can be distinguished. The resolution of an optical instrument is determined by either an instrument's pixel spacing or the diffraction-limited spatial resolution (Leach 2011). For LT sensors, there is a physical limitation due to speckle noise, which limits the achievable uncertainty (Dorsch *et al* 1994).

Figure 3.2 illustrates the height variations of the microfacets of a surface. Optically, a surface is classified as rough if its texture is characterised by dimensions (Δx, $h(x)$) that are large compared to the wavelength of an incident coherent illumination. Variations in heights result in scattered light with phase differences that can cause varying degrees of constructive and destructive interference. Interference of coherent and out-of-phase waves, where the path difference differs by close to or greater than the incident wavelength can result in the formation of bright and dark spots from coherent light scattering—this is known as 'speckle' (Goodman 1975). When the height variations are greater than one-fourth of the incident wavelength (Häusler *et al* 1999, Pavlicek and Hybl 2008, Hausler and

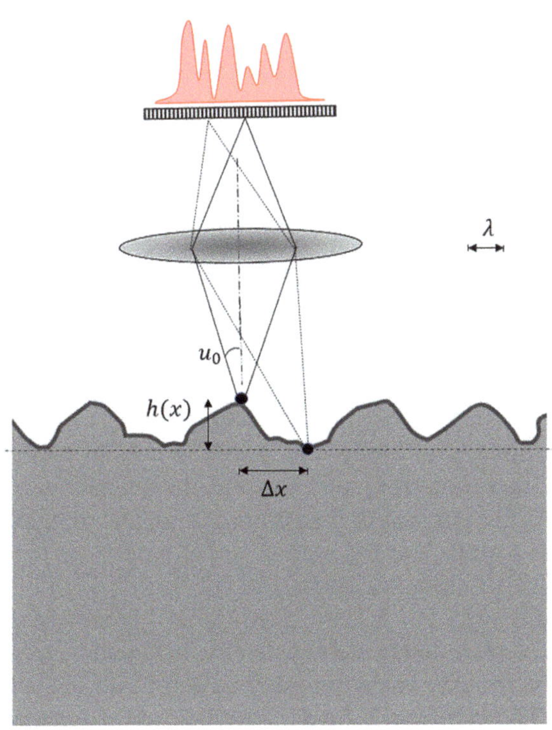

Figure 3.2. Speckle formation by scattering on a rough surface.

Ettl 2011), destructive interference may occur resulting in low contrast. Coherent light illuminated onto a rough surface, illustrated in figure 3.2, is scattered and results in an uneven distribution of light intensity.

When measuring the height variations in figure 3.2 along the laser spot beam in figure 3.1, $\delta h = \delta s$, the rate of change of the distance from point O, with respect to the lateral spot position can be expressed as

$$\delta h = \frac{1}{\sin(\beta)}\delta x. \tag{3.7}$$

The minimum lateral distance that can be resolved is limited by the spatial resolution. The spatial resolution (Leach 2011) of x is proportional to the wavelength of light and inversely proportional to the numerical aperture, $\sin(u_0)$,

$$\delta x = \kappa\frac{\lambda}{\sin(u_0)}, \tag{3.8}$$

where u_0 is the half-angle of the cone of light in figure 3.2. For a spot beam of coherent light, the lateral uncertainty can be derived by using speckle statistics and the value $\kappa = 1/2\pi$ was found in Dorsch *et al* (1994). The approach used by Dorsch *et al* was based on calculation of the standard deviation of the centre of gravity of laser spot intensity in a statistically derived speckle field. Using $\kappa = 1/2\pi$ in equations (3.7) and (3.8), the height variation can be obtained as

$$\delta h = \frac{1}{2\pi}\frac{\lambda}{\sin(\beta)\sin(u_0)}. \tag{3.9}$$

Equation (3.9) puts a theoretical limit to the achievable uncertainty of LT measurements. Increasing the observation aperture improves the uncertainty limit; however, it affects the measurement range negatively. The actual measurement capacity of LT sensors is frequently limited by other more dominant practical factors, such as lens imperfections, sensor noise and the geometric inhomogeneity of the measured surface.

3.3.2 Surface reflectance perspective

Surface topography at the micro- and macroscales affects the reflectance properties of a surface (Vukasinovic and Duhovnik 2019). When an incident electromagnetic wave interacts with the topography of a surface, the wavefront of the reflected beam will be distorted. In LT, a consistent distribution of detected light intensity is desired, but discontinuities in surface texture, colour and glossiness change the distribution in ways that are difficult to model (Curless 1997).

Reflectance models used for LT are commonly geometric models that provide approximations of the phenomenon described by physical optics. Geometric models have the benefit of simplicity over the more general and accurate physical models (Nayar *et al* 1991). The common models used to study surface reflection for LT are developed from the Beckmann–Spizzichino model (Beckmann and Spizzichino 1963) and the geometric Torrance–Sparrow model (Torrance and Sparrow 1966).

A geometric model, such as the Torrance–Sparrow, can be viewed from two geometric extremes—a sufficiently rough surface that allows light to scatter uniformly, and a mirror-like, perfectly smooth surface. The extreme case of uniform scattering does not exhibit specular behaviour and is governed by the Lambertian model of reflection. The Lambertian model describes the diffuse scattering of light that results in equal radiance from any observation angle. The luminance of the Lambertian reflection has a diffuse lobe, as shown in figure 3.3(a). When viewed from an observation angle of θ_r, the luminance is given by

$$I_{\mathrm{dl}} = K_{\mathrm{dl}} \cos(\theta_r), \tag{3.10}$$

where the maximum luminance at the perpendicular viewing direction is given by K_{dl}.

At the other extreme, a perfectly mirror-like object reflects laser light at a reflection angle that is equal to the incident angle. The concentrated, specular light spike at the fixed reflection angle is shown in figure 3.3(b). If K_{ss} is the spike luminance, the sharp spike of the light intensity observed at the reflection angle is modelled by delta functions in

$$I_{\mathrm{ss}} = K_{\mathrm{ss}}\delta(\theta_i - \theta_r)\delta(\phi_r), \tag{3.11}$$

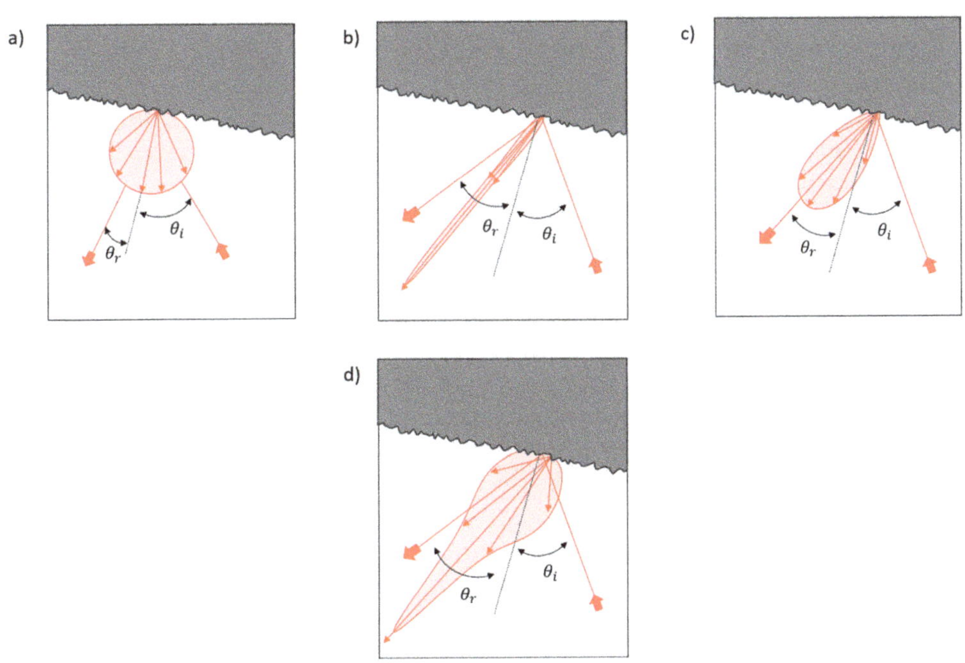

Figure 3.3. Luminance using reflectance models for (a) a Lambertian surface, (b) a specular spike for perfectly smooth mirror-like surface, (c) a specular lobe Torrance–Sparrow model and (d) a unified reflectance framework.

where ϕ_r is the geometrical attenuation factor which represents the proportion of light not attenuated by masking or shadowing. The Torrance–Sparrow model describes the specular lobe shown in figure 3.3(c); the derivation of the specular lobe I_{sl} can be found elsewhere (Nayar *et al* 1991). The specular luminance model describes surfaces that are between the Lambertian and mirror-like extremes. The specular lobe reflection is obtained by considering surfaces as a collection of light reflecting microfacets with light wavelengths much smaller than the texture of the surface (Nayar *et al* 1991).

From the three models of reflectance, a unified reflectance framework has been suggested where the reflectance of an object is considered as the superposition of the different models (Zhang and Li 2015, Nayar *et al* 1991). The superposed reflectance of an object, illustrated in figure 3.3(d), is then obtained by tuning the coefficients K_{dl} and K_{sl} and other parameters in the specular lobe model.

In LT, the distribution of observed reflected light is used to deduce information about an unknown surface. It is, therefore, easier to measure surfaces that exhibit more consistent luminance at different observation angles θ_r. Since the observed light intensities associated with the specular lobe and spike reflection models are highly dependent on θ_r, most reflection distributions in LT are assumed to have a Lambertian diffuse lobe (Nayar *et al* 1991) and specular reflections are avoided as much as possible. Practical camera and laser configurations are devised to mitigate specular spikes during measurement of simple objects whose surface normal directions can be roughly estimated. The configurations, illustrated in figure 3.4, are grouped based on the triangulation geometries formed by the camera, laser and measured surface (Stavroulakis and Leach 2016, Latimer 2015).

The four triangulation geometries are suitable for different surface properties and applications (Latimer 2015):

- The standard geometry is computationally simple and is used for general applications without severe specular reflections.
- The reverse geometry is also suitable for general applications without severe specular reflection. It takes more computation than the standard configuration but provides higher accuracy.
- Specular geometric configurations can cause errors on most metallic surfaces; however, for surfaces with dark coloured texture, the specular configuration can be used to increase the intensity of received light.
- To prevent specular reflection on shiny objects, the look-away configuration can be used. This configuration keeps the camera viewpoint at the opposing side of the specular lobe peak.

Specular and spike reflections may not be easy to avoid for many intricate shapes and the geometric configurations could vary at different positions of an object. For concave surface regions, the laser beam can undergo secondary and higher order reflections that make image feature extraction difficult and even impossible in some cases. Figure 3.5 shows the higher order reflections, where distinguishing the primary reflected beam from the other reflections can be difficult.

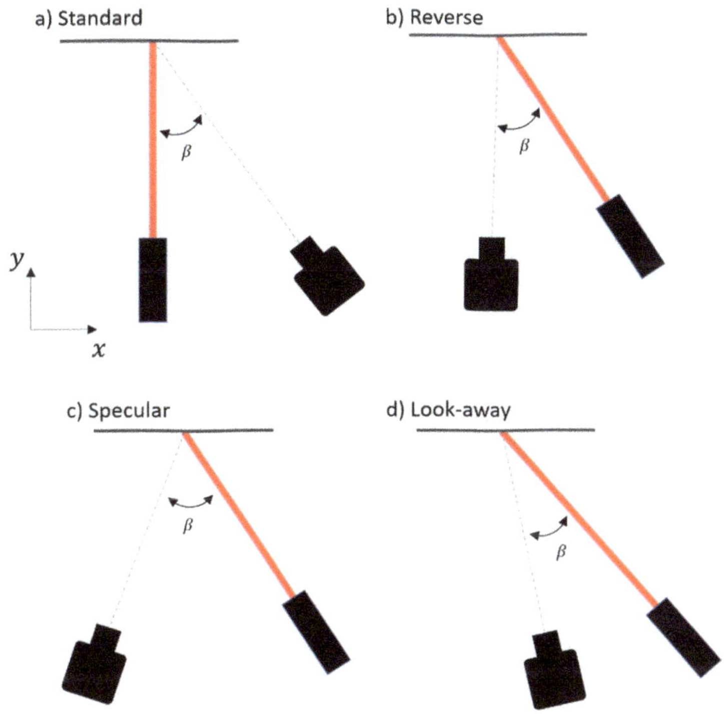

Figure 3.4. The four geometric configurations of laser plane, camera and surface: (a) standard, (b) reverse, (c) specular and (d) look-away.

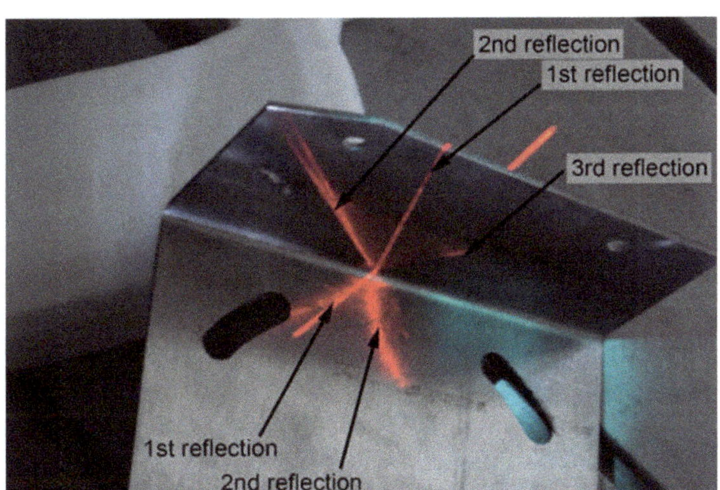

Figure 3.5. Demonstration of secondary and higher order reflections (Vukasinovic and Duhovnik 2019 with permission of Springer).

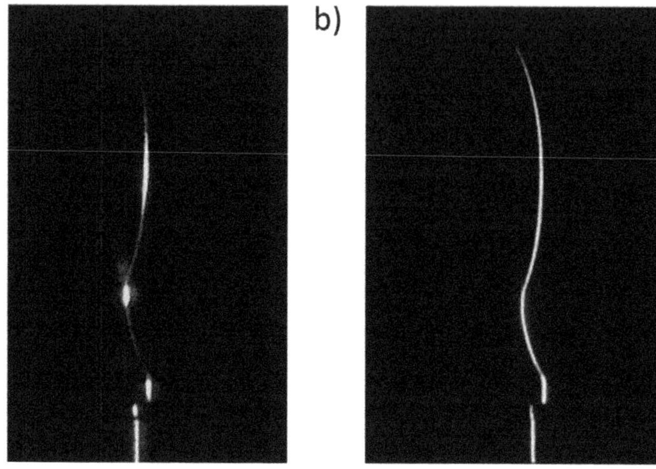

Figure 3.6. The use of coating to reduce specular reflections (reprinted from Isa and Lazoglu 2017 with permission of Elsevier).

Higher order reflections can cause significant measurement error (Vukasinovic and Duhovnik 2019); hence, smart algorithms are often used to discriminate higher order reflections and mitigate their impact (Amir and Thörnberg 2017). The probability of occurrence of the reflections increases with the shininess and concavity of the measured surface. Shiny surfaces can exhibit specular reflections that contribute to higher order reflections. A practical method for decreasing specular reflection is by using matte spray paints or coatings (Pereira *et al* 2019). The impact of coating on an aluminium sample to reduce specular reflection is demonstrated in figure 3.6 (Isa and Lazoglu 2017).

3.3.3 Measurement dependence on surface form

From the reflectance models discussed in section 3.3.2, the incident and observation angles are required. However, for many freeform objects, the angles can be difficult to estimate without knowledge of the surface form. Hence, the measurement of surface form from images of laser light is a difficult problem because the reflected light is itself dependent on the form of the surface.

Zhou *et al* (1998) showed that the luminous power received from a surface and the detected centroidal position of a laser beam are dependent on the surface form. When a thin sheet of laser light is reflected from the surface of a measured object, as shown in figure 3.7, the reflected light is received by the camera through the solid angle Ω of the receiving lens with radius r_l. The projected light is assumed to obey Lambert's law which assumes the light is diffuse and its luminance is independent of the viewing angle. With the further assumption that the power received in the lens is evenly distributed, the received scattered light power can be expressed as the product of the solid angle and the intensity of the received light, $P = I\Omega$. For a Lambertian reflection, the intensity of the reflected light is given by

$$P = I_0 \frac{\pi r_l^2 \cos(\beta - \gamma)}{(z_0 - h\cos(\beta))^2}, \tag{3.12}$$

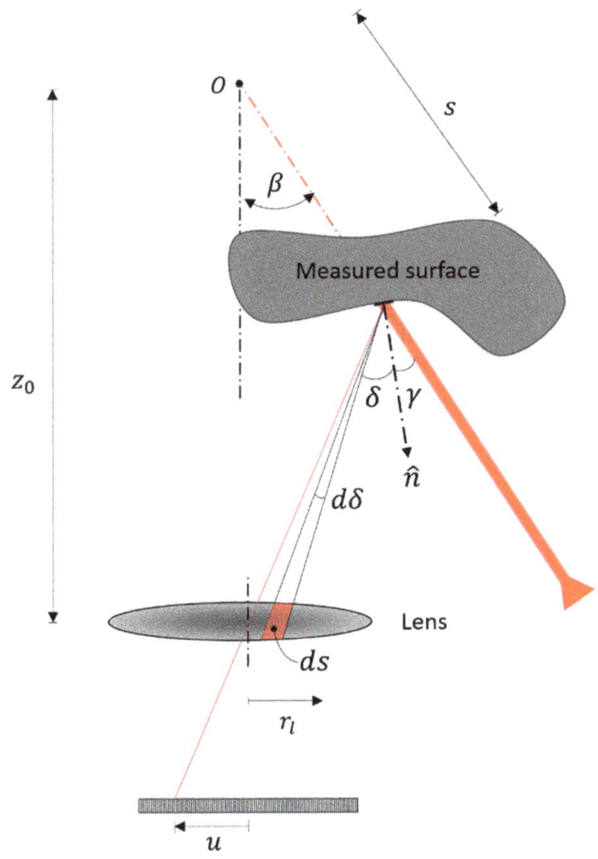

Figure 3.7. Segment of observed light from an inclined surface.

assuming r_l is small compared to the distance of the projected light from the centre of the lens and γ is the inclination angle of the measured surface (Zhou *et al* 1998). Equation (3.12) shows that the laser power received by the lens is dependent on the surface form parameters γ and h.

By integrating the Lambertian intensity of the laser light over the observation solid angle of the lens, due to the differential surface ds at angle δ shown in figure 3.7, the angle δ_m where the received power is halved is given by

$$\delta_m = \frac{r_l^2}{z_0^2}\left(1 + 2\frac{x}{z_0}\cos(\beta)\right)\tan(\beta - \gamma). \tag{3.13}$$

Using equation (3.13), the form error, which is the deviation of the centre of gravity of the laser light along the distance s, is given by (Zhou *et al* 1998)

$$\Delta s = \frac{l_0 r_l \tan\gamma}{z_0 \tan\beta}\left(1 - \frac{s\cos\beta}{z_0}\right)\left(1 + \frac{2s\cos\beta}{z_0}\right). \tag{3.14}$$

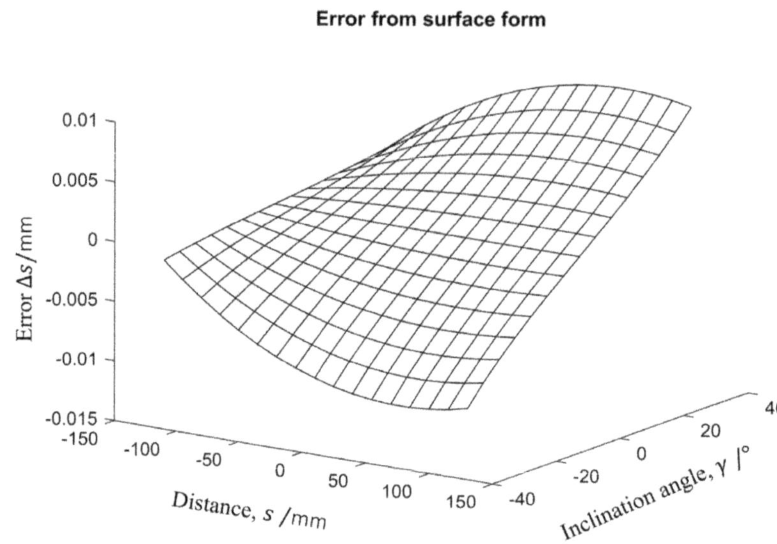

Figure 3.8. Plot of sample error caused by surface form parameters.

The form error is also referred to as tilt or inclination error because the inclination angle γ has a significant impact on the derived error (Li *et al* 2016, Sun and Li 2016). Using equation (3.14), for $l_0 = 0.3$ mm, $r_l = 8$ mm, $z_0 = 190$ mm and $\beta = 45°$, the form error is plotted in figure 3.8.

Form error can dominate the speckle error and methods of compensation are needed for accurate measurements (Ding *et al* 2020). Some recommendations to decrease inclination error include the use of additional cameras (Zhou *et al* 1998) and adjusting the orientation and position of the LT sensor (Li *et al* 2014).

3.4 Laser triangulation systems

LT systems consist of sensor units, mounting components and motion systems for positioning objects when carrying out measurements. LT systems can only measure the external surfaces of objects and cannot measure volumetric and internal features. Depending on the application, LT systems can be designed to carry out various types of measurement, from 1D distance measurement to 3D point clouds. For coordinate metrology, LT systems commonly use a combination of LT sensor units and various forms of scanning systems. This section covers the extension of point based LT sensor and integration of scanning systems in different LT systems.

3.4.1 Extension of a point based laser triangulation sensor

The 1D LT sensor discussed in section 3.2 can be extended to 2D or 3D by shaping the laser beam into one or more lines. It is common to use line generating optics with semiconductor laser diodes to construct line lasers. The aim of the laser line generation process is to transform a focused spot laser beam to a line beam of uniform intensity along the line length and a Gaussian distribution across the width

(Craggs *et al* 2012). Typically, line laser generators for LT are constructed by either the arrangement of lenses for beam stretching or the incorporation of rotating micromirrors. Conventional lenses for line generation include cylindrical and Powell lenses (Gruber *et al* 2018), which both have their benefits and drawbacks. While Powell lenses provide more uniform line beam intensity than cylindrical lenses, they require optimisation for different line beam specifications and are less adaptable. Additional optical components can be used to improve the uniformity of cylindrical lenses (Lin *et al* 2016). A tandem lens array or specially designed lens can also be used to generate line laser beams (Craggs *et al* 2012, Wang *et al* 2014). The other strategy for laser line generation involves the use of vibrating micromirrors at high frequency (Zuo *et al* 2017, Zuo and He 2018). The vibrating mirror strategy can generate more uniform illumination but lacks robustness due to the necessary moving components.

Figure 3.9 shows the increased dimensions of measurements obtainable by LT using some combinations of laser line beams. While the 1D LT sensor is suited to distance measurement, it will be too slow for profile and surface measurement; therefore, 2D LT sensors can be employed to achieve faster measurement rates per image. To further increase the rate of measurement per image, it is possible for LT sensors to project multiple line laser beams. Although using multiple laser lines makes the extraction of points from images more difficult, the use of more complex structured light projection has given rise to a 3D coordinate measurement method referred to as fringe projection (covered in detail in chapter 5 of this book).

Even though an LT can be extended to carry out 2D and 3D measurements, in practice, a mechanism for the movement of the LT sensor or the measured object is necessary. The motion system ensures that desired points or regions on the measured objects are illuminated and detected by the LT sensor. It is, therefore, important to investigate suitable combinations of the various types of LT sensor and motion systems for different coordinate measurement applications.

For the remainder of this section, LT systems are classified based on the dimension of the desired measurand obtainable. Hence, LT systems are grouped based on the nature of what they measure. Drouin and Beraldin (2012) used a similar categorisation: spot, stripe and area systems. The simplest are the point LT systems that measure either the coordinates of a point or a distance. Following the

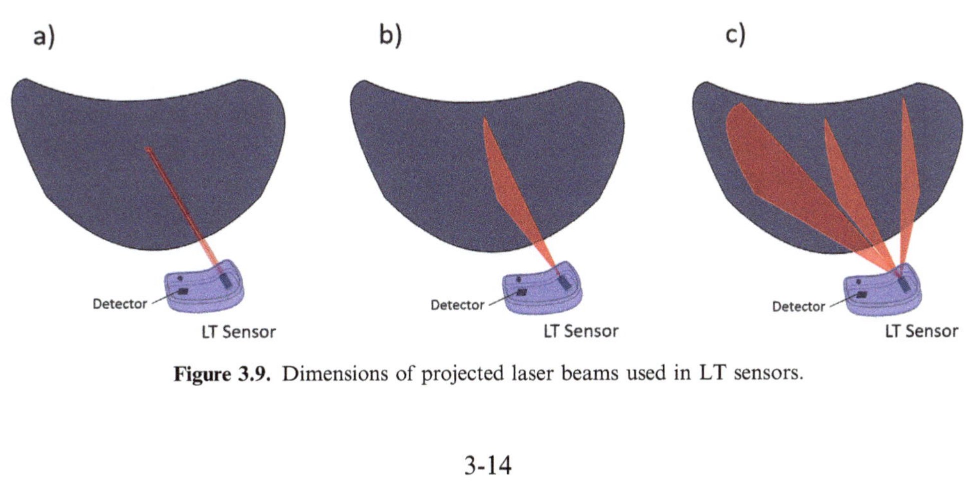

Figure 3.9. Dimensions of projected laser beams used in LT sensors.

point systems, profile LT systems measure 2D features, such as lines and circles. Finally, surface LT systems are capable of measuring 3D surface forms. This classification makes it possible to study LT systems for distance, profile and surface measurements separately, and analyse the strategies used for each.

3.4.2 Point measurement systems

The fundamental principle of LT was presented in section 3.2. From early developments, LT systems were used as convenient non-contact distance measurement sensors using spot laser beams (Ji and Leu 1989, Okada 1982). Despite several variations, distance measuring LT systems are conventionally designed to conduct distance measurement along the laser beam direction. The wavelengths of the laser radiation are commonly within the visible light spectrum, therefore, providing an auxiliary function as a visual mark of the measured point on a surface. In addition to distances, as described in section 3.2, the lateral and depth coordinate positions can also be measured. From figure 3.1, the distance of the laser spot from the reference point O can be evaluated, using equations (3.3) and (3.4) as

$$s = \frac{x}{\sin(\beta)} = \frac{u}{f_x \sin(\beta) + u\cos(\beta)} z_0. \tag{3.15}$$

The sensitivity of the detected spot position u with respect to the displacement s can be expressed as

$$\frac{\partial u}{\partial s} = \frac{f_x \sin(\beta) + u\cos(\beta)}{z}. \tag{3.16}$$

The sensitivity to the displacement along the laser line, given in equation (3.16), is inversely proportional to the measurement depth. Increasing the focal length of the lens can increase the sensitivity at the cost of reducing the field of view. Though limited by manufacturing capability and cost, the sensor size and pixel density can be increased to improve the sensitivity.

The distance measured, as given by equation (3.15), is based on a simplified model of the focusing lens, where received laser light is assumed to be accurately focused at a point. When reflected light is received through a real objective lens, the laser spot is focused on the objective-aligned detector at variable degrees of sharpness along the working range. Decreased sharpness of the observed light spot amounts to higher uncertainty in the position of the spot. The Scheimpflug principle, or Scheimpflug condition, is commonly applied to tilt the detector plane to maintain uniform focus of the detected laser light (Gao *et al* 2018). The Scheimpflug condition is satisfied when the detector plane, the lens plane and the laser line intersect at the point L in figure 3.10. The intersection occurs when the angle between the detector and the optical axis is $\beta_d = \tan^{-1}(\frac{z_0}{f}\tan(\beta))$. The distance of the laser beam in the tilted Scheimpflug condition can be derived as

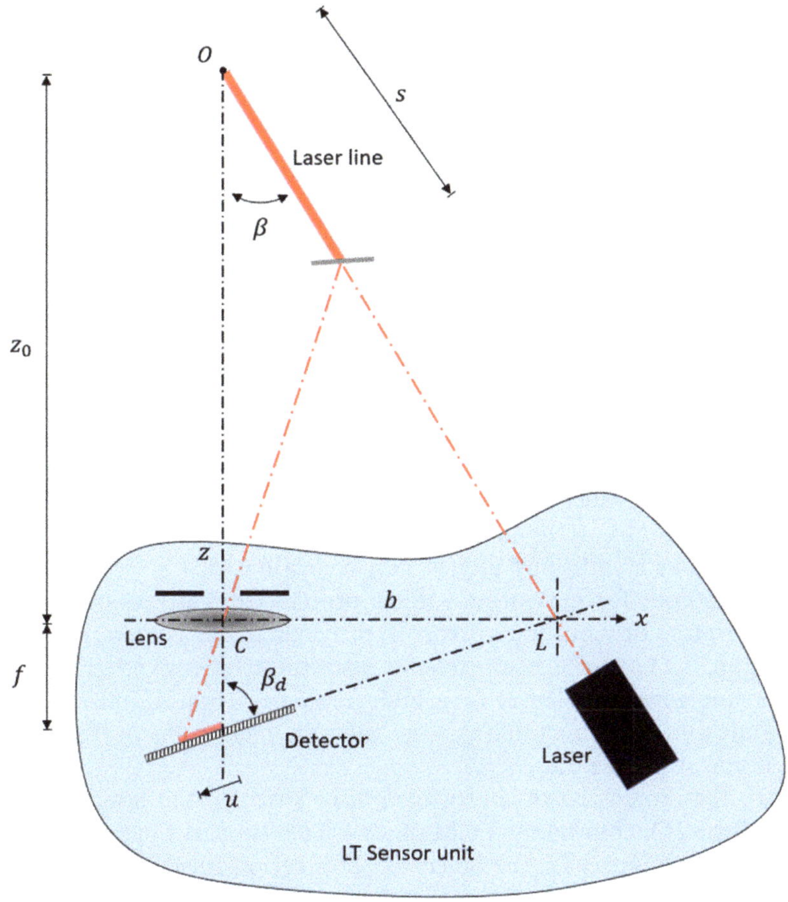

Figure 3.10. Scheimpflug condition showing the intersection of laser line, detector and lens plane.

$$s = \frac{u\sin(\beta_d)}{f_x \sin(\beta) + u\sin(\beta + \beta_d)} z_0. \quad (3.17)$$

The parameters β, β_d, z_0 and f_x in equation (3.17) are chosen based on the design requirements of the LT sensor, such as measurement range, accuracy and cost.

The tilted configuration of the detection plane is commonly used in displacement sensors to extend the depth of field of the LT sensor by maintaining in-focus detection. However, the Scheimpflug condition can be complex to implement when line laser illumination is used (Peterson and Peterson 2006, Schlarp *et al* 2020). Furthermore, it is impossible to fulfil the intersection condition of the Scheimpflug condition when multi-line or grid type illumination is used. Just as the effect of surface form on measurement uncertainty was analysed for a non-tilted detector in section 3.3, the inclination error in the tilted configuration has been studied (Dong *et al* 2018).

3.4.3 Profile measurement systems

The cross-sectional profile of a surface can be measured using LT profile measurement systems. Measured profile data using LT normally represents a best-fit or average of the actual profile, which filters out microscale surface texture information. In principle, microscale features can be measured at close range using thin beam widths, but the depth of focus becomes narrow and the uncertainty influence factors discussed in section 3.3 become more significant. Therefore, major industrial applications of LT systems recognise it as a form (rather than a texture) measurement technique.

A simple profile measuring system can be devised by attaching a 1D LT sensor to a one degree of freedom motion system. The speed of measurement of the point-wise measuring system is low because a surface profile is measured one point at a time. Therefore, 2D LT sensors that measure one stripe at a time, are commonly employed for profile measurement. From a stripe measured from the surface of an object, 2D features, such as lines, triangles, ellipses and freeform curves, can be measured (Carmignato *et al* 2020). Using the measured feature data, critical dimensions, such as lengths and angles, can be evaluated.

Prevalent methods for measuring surface profiles from stripes involve the use of laser line generators or vibrating mirrors to project a linear beam onto a surface. As shown in figure 3.11, a spot beam from a laser diode is converted to a line beam using either line generator lenses or a high-frequency rotating mirror. The mirror needs to rotate at a frequency that is more than the measurement frequency used to acquire LT sensor profile data.

Similar to the dimension of a projected laser stripe in the laser plane, the light detector must be 2D, commonly a CMOS or CCD (Isa and Lazoglu 2017, Molleda *et al* 2013). From figure 3.12, the detected 2D pixel position is mapped from a 3D

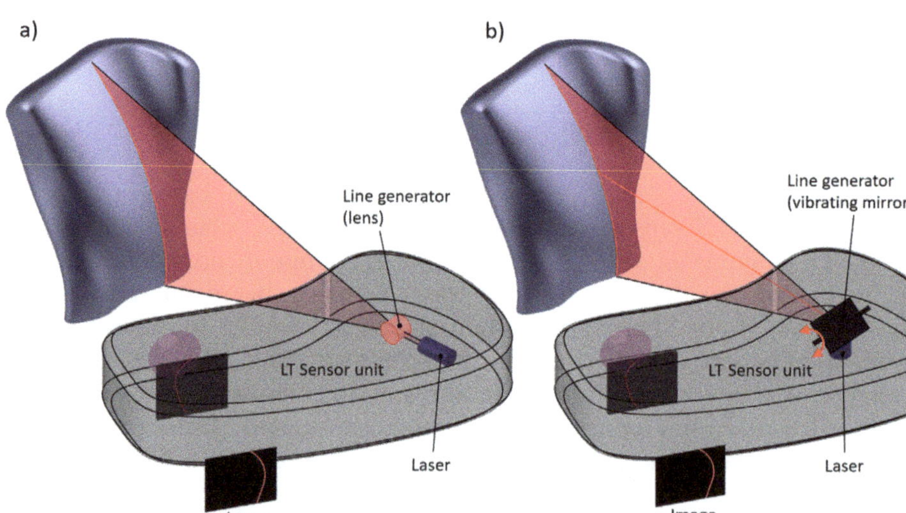

Figure 3.11. Line generators used in LT systems.

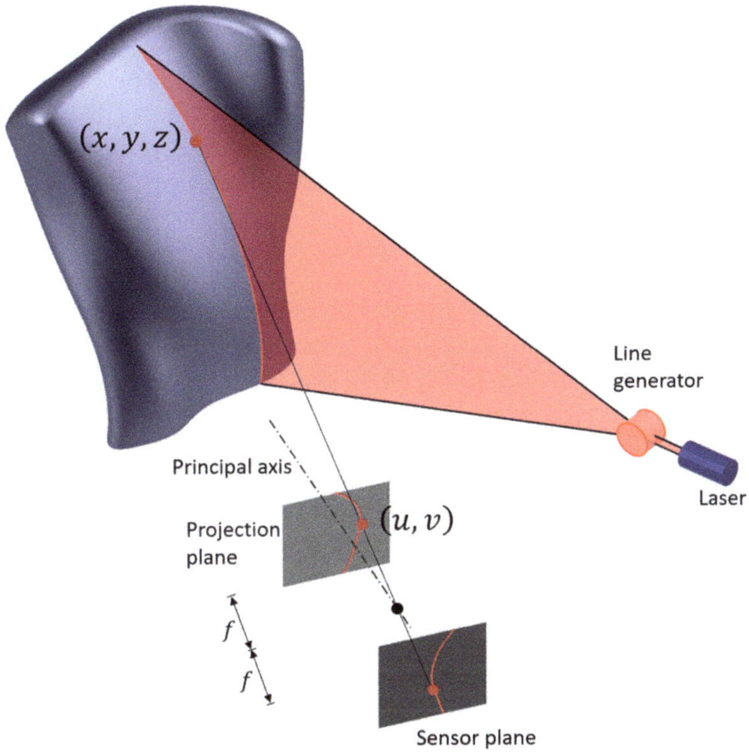

Figure 3.12. Triangulation of laser stripe points (x, y, z), image projection (u, v) and the laser.

point of the laser stripe. Therefore, in addition to equations (3.3) and (3.4), a third expression for y is necessary. Where v and f_y are the vertical pixel position and the vertical focal length, respectively, the 3D coordinate of a point located at the image position (u, v) can be expressed as

$$\begin{bmatrix} x \\ y \\ z \end{bmatrix} = \begin{bmatrix} \dfrac{u\tan(\beta)}{f_x\tan(\beta) + u} z_0 \\ \dfrac{vf_x\tan(\beta)}{f_x f_y\tan(\beta) + f_y u} z_0 \\ \dfrac{f_x\tan(\beta)}{f_x\tan(\beta) + u} z_0 \end{bmatrix}. \tag{3.18}$$

The pixel position (u, v) is measured from the principal axis on the image. The central projection model places a virtual projection plane at an opposite and equal distance to the image sensor from the projection centre. The image pixel positions are represented on the projections plane in figure 3.12.

The terms 'laser light section' and 'light sheet triangulation' have also been used to describe the laser stripe measurement method because the measured surface profile results from the intersection of an object and a laser plane (Donges and Noll 2015, Schwenke *et al* 2002). Using the known geometric relationship between the camera and laser light, 3D coordinates can be determined. The geometric relationship used to obtain equation (3.18) is for a vertical plane of laser light. Similar triangulation relations for horizontal and slant laser line projections can also be derived to obtain the measured 3D points (Idrobo-Pizo *et al* 2019).

3.4.4 Surface measurement systems

When optical systems are confined to measure only points and profiles, there can be a high probability of missing significant features on a surface (Leach 2011). A representation of a surface beyond profiles and points is necessary to address the possible omission of important detail. The requirement for a high number of points, the feasibility of faster computational speeds and automation of measurement procedures has increased the interest in areal surface LT measurement. Compared to contact CMSs that measure at scanning speeds below 10 mm s^{-1} and gather a maximum of few hundred points/second, LT systems can measure up to a million points per second over relatively large surface areas (Matharu *et al* 2019, Ghiotti *et al* 2015).

In the same way that the use of 1D LT sensors is not preferred for profile measurement, the combination of a 1D LT sensor and a 2D scanning mechanism for surface measurement is rarely used because of low speeds. Therefore, point-wise surface measurement systems are not considered here. Figure 3.13 shows the possible scanning mechanisms that are commonly integrated with 2D or 3D LT sensors. In figures 3.13(a) and (b), the respective 1D translational and rotational motion systems are used to scan the surface of an object. Alternatively, the object can be translated or rotated depending on the type of application and the size of the object. Figure 3.13(c) shows the case where a rotating mirror is used to scan the object. There is a distinction between the rotating mirror in figure 3.13(c) and the vibrating mirror used for laser line generation in section 3.4.2. The rotating mirror in figure 3.13(c) moves the laser plane across the object surface as the laser stripe

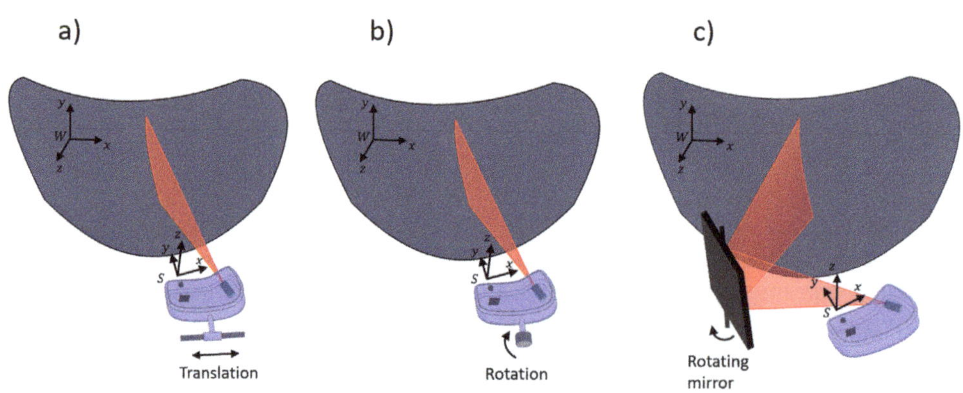

Figure 3.13. Scanning mechanisms for surface LT measurement.

images are captured. However, the line generating mirror distributes the light luminosity across the profile of the object before an image is taken. In this chapter, the vibrating mirror is considered to be part of the LT sensor as an alternative laser line generating module.

For the translational and rotational scanning modes, the relative displacements of the LT sensor coordinate system S with respect to the fixed coordinate W can be expressed with a homogeneous transformation matrix ${}^{W}\boldsymbol{T}_S$. The LT sensor coordinate system is chosen as the imaging projection centre. Beyond the 1D motion, the relative motion of a scanning mechanism up to the maximum six degrees of freedom can be encapsulated in the transformation matrix and the 3D coordinate position of the points can be given as

$$\begin{bmatrix} x \\ y \\ z \\ 1 \end{bmatrix} = {}^{W}T_S \cdot \begin{bmatrix} \dfrac{u\tan(\beta)}{f_x\tan(\beta) + u} z_0 \\ \dfrac{vf_x\tan(\beta)}{f_xf_y\tan(\beta) + f_y u} z_0 \\ \dfrac{f_x\tan(\beta)}{f_x\tan(\beta) + u} z_0 \\ 1 \end{bmatrix}. \tag{3.19}$$

As for the case where a mirror scans the object, the triangulation geometry varies when the mirror rotates; in particular, the values of the triangulation angle β and the reference depth $z_0 = b/\tan(\beta)$ in equation (3.19) change during measurement. For simplicity, the mirror can be positioned to keep the baseline distance b fixed. Then, with a variable triangulation angle β used for scanning, the 3D coordinates of the points of the laser stripe are given by

$$\begin{bmatrix} x \\ y \\ z \end{bmatrix} = \begin{bmatrix} \dfrac{u}{f_x\tan(\beta) + u} b \\ \dfrac{vf_x}{f_xf_y\tan(\beta) + f_y u} b \\ \dfrac{f_x}{f_x\tan(\beta) + u} b \end{bmatrix}. \tag{3.20}$$

Multiple laser lines are used to increase the measurement speed and can be used with any of the scanning methods shown in figure 3.13, and more flexible methods covered later in this section. However, when relying solely on the LT relations, distinguishing the laser lines and extracting accurate pixel locations from images can be a challenging task. The problem is compounded by practical error sources, such as secondary reflection, the presence of occlusions and variable material properties (Curless 1997).

3.4.5 Advanced laser triangulation systems

Measurements by LT systems are affected by triangulation error of the sensor and errors in positioning the LT sensor unit. These errors are inevitable in LT (Yao *et al* 2019), and statistical filtering might not be enough when there are numerous points with large errors. As a result, strategies that provide supplementary correlation, in addition to the triangulation relations covered in the previous parts of section 3.4, have been introduced to improve the accuracy of measurements.

It is common to use additional camera(s), where passive (photogrammetric) triangulation relations are supplemented with active triangulation. Stereo vision, where two cameras are used, adds a correlation between the corresponding pixel features through an epipolar constraint, covered in chapter 4. The stereo vision relationship can be used to identify corresponding laser points in two cameras. The use of two (or more) cameras can reduce error in the identification of laser locations in images and reduce inclination error (Zhou *et al* 1998, Li *et al* 2016). Furthermore, relying on passive triangulation and multi-view correspondence can make it easier for flexible scanning mechanisms to be used. Using the term 'passive triangulation' in this context is confusing because the overall measurement is active, however, the triangulation is passive because it does not require the geometric information of the laser beam. Hence, without the need to characterise the geometric location of the laser source, stereo and multi-view correspondences assist in making flexible LT system designs possible.

An optical modification to the conventional LT range sensor has been proposed to enable expansion of the field of view (FOV) without compromising resolution (Rioux 1984, Johannes *et al* 2018). The scanning rotation of the projected illumination is synchronised with the change in the optical path of the received light before reaching the detector. In figure 3.14, for equivalent focal length and geometric variables, it can be seen that the synchronised system occupies a smaller

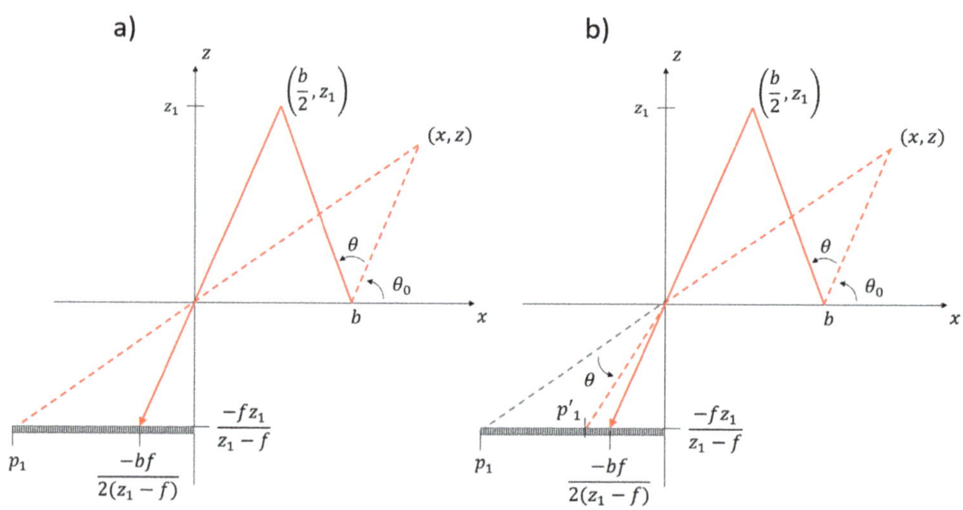

Figure 3.14. Synchronised scanning LT systems compared with conventional systems.

FOV. This shows that, for similar geometric scales, the synchronised system permits the use of lenses with larger focal lengths without reducing depth resolution.

An auto-synchronisation scanning system shown in figure 3.15 shows a realisation of the synchronised scanning strategy using a polygon mirror (Zhang *et al* 2014). To ensure high speed in profile measurement, the mirror rotation is encoded to allow acquisition of laser lines.

Other realisations of the idea of synchronised scanning exist (Tu *et al* 2019), as well as other techniques to increase the FOV using multi-view optics (Li *et al* 2018). Industrial on-site use of LT favours hand-held and robot-arm mounted LT sensors to enable adaptable path planning over wider areas. For these systems, tracking the position of the LT sensor during measurement is critical. Recently, as alternatives to costly laser trackers, photogrammetry-based trackers employing cooperative targets have been used to measure the position of the LT sensor (Sun *et al* 2017). The localisation of the LT sensor is commonly carried out by stationary or LT sensor-attached multi-view systems. Resolving the space and time multi-view correspondence (Song 2013) is critical for accurate tracking and 3D measurement.

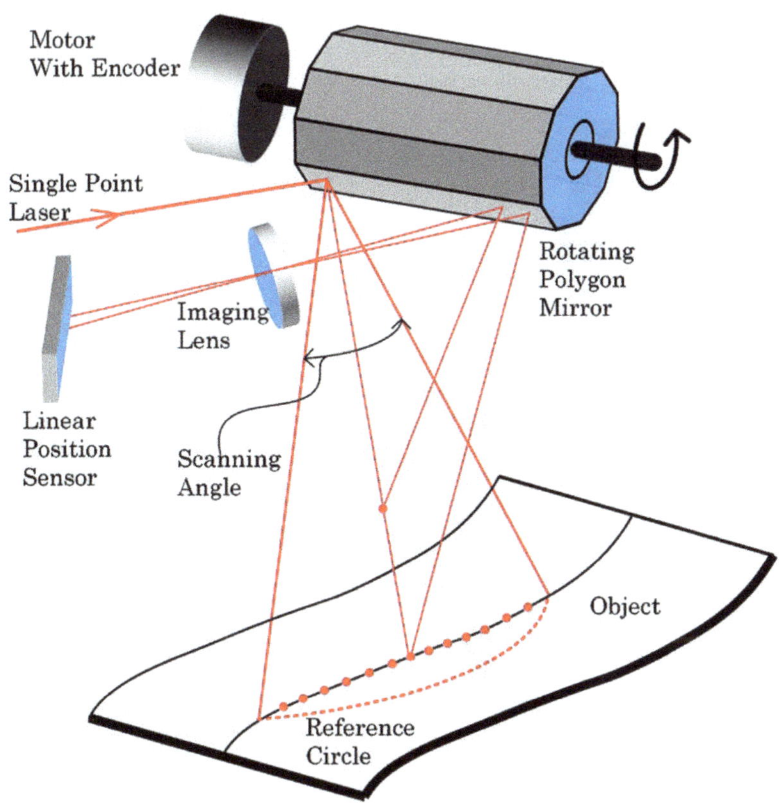

Figure 3.15. Realisation of a synchronised LT system (reprinted with permission from Zhang *et al* 2014, copyright The Optical Society).

Current LT measurement systems demand high flexibility to allow measurement of more industrial products, where geometric complexities of parts are on the rise (Raghavendra *et al* 2020). The flexibility offered by robot and hand controlled LT sensors is desirable; however, accurate LT sensor extrinsic parameter measurements, using devices such as trackers, articulated arm CMSs and contact CMSs, are necessary (Sousa *et al* 2017). Figure 3.16 shows some of the advanced LT sensor systems that are currently used in industry. LT probes are commercially sold for use on tabletop contact CMSs where the tactile probes are replaced by LT sensors as illustrated in figure 3.16(a). Figure 3.16(b) shows a hand-held LT sensor attached to an articulated arm to register the position and orientation of the sensor. Tracker and fixed markers are used in figures 3.16(c) and (d), respectively, to monitor flexible LT sensor positions.

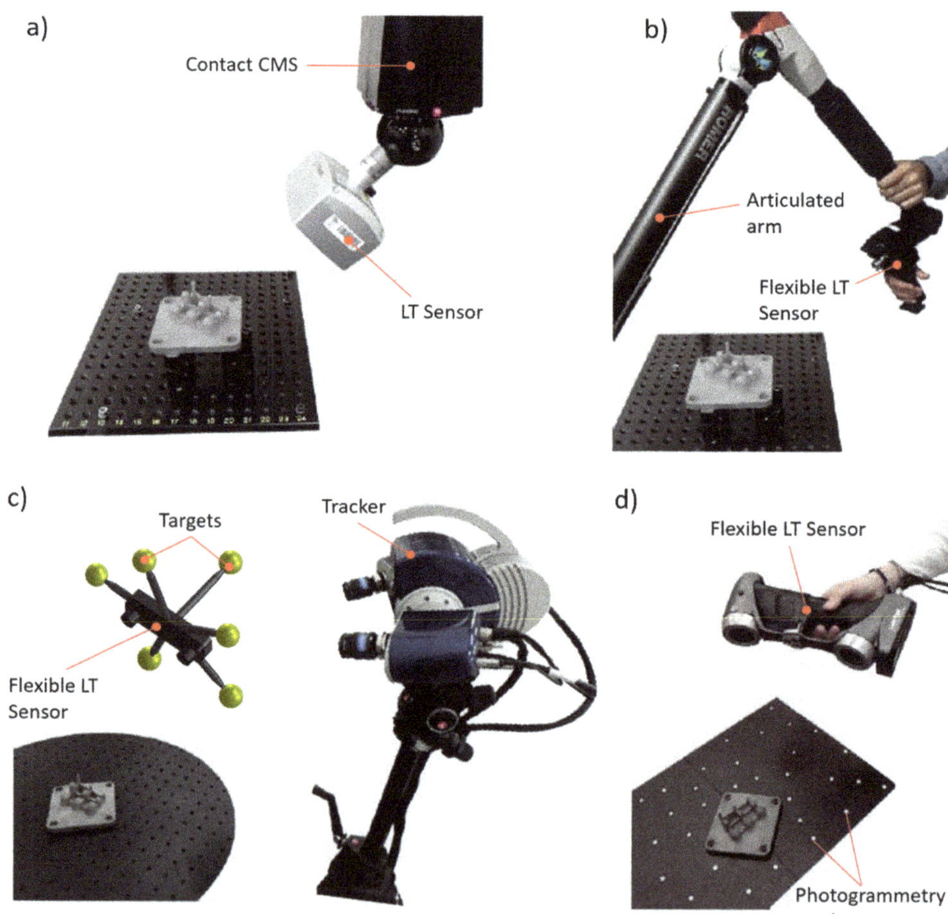

Figure 3.16. Advanced LT systems using (a) a contact CMS, (b) an articulated arm, (c) a tracker and (d) photogrammetry, for positioning of LT sensors. Adapted from Giganto *et al* (2020) with permission from MDPI.

3.5 Working process of laser triangulation

This section covers the general processes involved in coordinate measurement using LT systems. The measurement principle of LT systems discussed in section 3.4 relies on processes, such as image acquisition, determination of imaging parameters, LT sensor localisation and extraction of pixel position from images. The quality of LT measurements depends on the current technological developments in these working processes and future improvement of the LT systems rely on innovations in the processes.

Figure 3.17 shows the flowchart of major processes involved in measurement of objects by LT systems. It includes stages for the determination of system parameters that are needed to acquire information from images. Figure 3.17 also delineates the procedures for extraction of images pixel locations and how they are used to reconstruct 3D points.

3.5.1 Characterisation of intrinsic parameters

Before measurements are carried out, there is the need to identify how the images captured relate to the real object they represent. Identified pixel locations on image sensors need to be correlated with the metric scale from the 3D scene. To accomplish this, a process of characterisation of the internal (intrinsic) parameters of the observation optics is necessary. Based on the central projection imaging model, the intrinsic parameters are the focal length of the lens, the pixel coordinates of the principal axis, the distortion coefficients and skew coefficient.

The radial and tangential distortion coefficients (k_1, k_2, … and p_1, p_2…, respectively) are commonly obtained from the truncation of the model given by

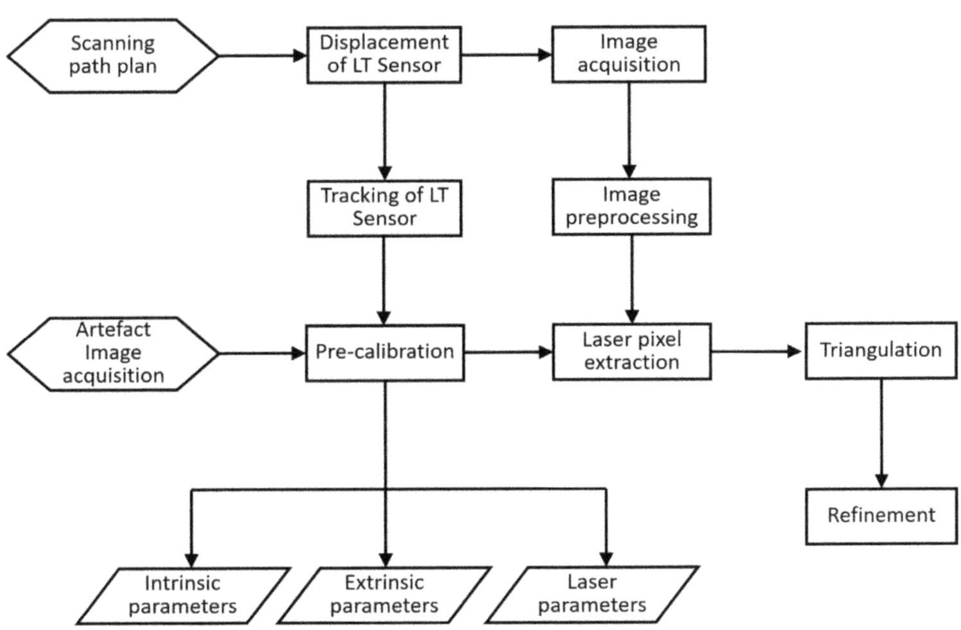

Figure 3.17. Flowchart of processes involved in LT.

$$\begin{bmatrix} \tilde{u} \\ \tilde{v} \end{bmatrix} = \begin{bmatrix} u + u\sum_{i=1}^{n} k_i\rho^{2i} + (p_1\rho^2 + 2p_1u^2 + 2p_2uv)\left(1 + \sum_{i=1}^{n} p_{2+i}\rho^{2i}\right) \\ v + v\sum_{i=1}^{n} k_i\rho^{2i} + (p_2\rho^2 + 2p_1v^2 + 2p_1uv)\left(1 + \sum_{i=1}^{n} p_{2+i}\rho^{2i}\right) \end{bmatrix}, \quad (3.21)$$

where $\rho = \sqrt{u^2 + v^2}$ (Zhang 2000, Drap and Lefèvre 2016). The image coordinates are defined with respect to the principal point, therefore, both the distorted $(\tilde{u}, \tilde{v})$ and undistorted pixel positions (u, v) are dependent on the principal reference point (u_0, v_0). With the exception of wide-angle lenses, a truncation of $n = 2$ or $n = 3$ is usually satisfactory for radial distortion, while tangential distortions are negligible in many applications (Isa and Lazoglu 2017).

3.5.2 Characterisation of extrinsic parameters

The relationship between the LT sensor coordinate (S) and a world coordinate frame (A) is necessary for LT system characterisation and measurement. The parameters describing the position and orientation of the LT sensor with respect to the world coordinate system are referred to as the extrinsic parameters. These parameters describe the pose of the camera and are usually needed during measurement or pre-calibration of LT systems. The extrinsic parameters can be expressed through rigid body transformations. Although, general rigid body transformations are described by six independent parameters, the convenient homogeneous transformation matrix has nine non-zero elements (i.e. the elements are not independent). Six elements describe the orientation of S with respect to A and three elements describe the relative position.

3.5.3 Pre-calibration of laser triangulation systems

The process of determining the intrinsic and extrinsic parameters may vary for different LT systems but it is an essential procedure. The triangulation equations (3.18)–(3.20) show that the coordinate positions depend on various geometric variables and need to be determined accurately. These variables are associated with the camera, the laser or the relative pose of the LT sensor with respect to the measured object.

For a camera with unknown intrinsic and extrinsic parameters, the mapping from 3D points (x, y, z) to 2D image pixels $(u + u_0, v + v_0)$ can be expressed as

$$\tilde{w}\begin{bmatrix} u + u_0 \\ v + u_0 \\ 1 \end{bmatrix} = [\boldsymbol{I}_{3\times3} \;\; \boldsymbol{0}_{3\times1}] \cdot \begin{bmatrix} f_x & s & u_0 & 0 \\ 0 & f_y & v_0 & 0 \\ 0 & 0 & 1 & 0 \\ 0 & 0 & 0 & 1 \end{bmatrix} \cdot {}^{A}\boldsymbol{T}_S^{-1} \cdot \begin{bmatrix} x \\ y \\ z \\ 1 \end{bmatrix}. \quad (3.22)$$

Because the image principal point (u_0, v_0) is unknown, the left-hand side of equation (3.22) uses pixels $(u + u_0, v + v_0)$ measured from the lower left corner of the image projection plane. The identity matrix $\boldsymbol{I}_{3\times3}$ and zero vector $\boldsymbol{0}_{3\times1}$ make up the dimensionality reduction matrix $[\boldsymbol{I}_{3\times3}\ \ \boldsymbol{0}_{3\times1}]$. The skew parameter s is commonly zero, $\tilde{w}$ is the scaling factor and the extrinsic parameters are given by ${}^{A}\boldsymbol{T}_{S}$. The camera matrix $\boldsymbol{P}$, extrinsic matrix ${}^{A}\boldsymbol{T}_{S}$ and intrinsic matrix $\boldsymbol{K}$ are related through

$$\boldsymbol{P} = \boldsymbol{K} \cdot {}^{A}\boldsymbol{T}_{S}^{-1} = \begin{bmatrix} f_x & s & u_0 & 0 \\ 0 & f_y & v_0 & 0 \\ 0 & 0 & 1 & 0 \end{bmatrix} \cdot {}^{A}\boldsymbol{T}_{S}^{-1}. \tag{3.23}$$

Using the projection matrix in equation (3.23), equation (3.22) can be rewritten as

$$\tilde{w}\begin{bmatrix} u + u_0 \\ v + u_0 \\ 1 \end{bmatrix} = \boldsymbol{P} \cdot \begin{bmatrix} x \\ y \\ z \\ 1 \end{bmatrix}. \tag{3.24}$$

A popular method for determining camera intrinsic parameters is by using an artefact with a calibrated dimension. The artefact must have feature points that can be accurately identified and easily matched in images. For determination of intrinsic parameters, it is common to use planar targets to reduce the camera matrix to a homography matrix by aligning the origin and xy-plane of the coordinate system A to the artefact plane (Isa and Lazoglu 2017, Hartley and Zisserman 2003). The images of the target are normally captured at various locations and poses to determine the intrinsic parameters. This process is only useful for the determination of intrinsic parameters because the extrinsic parameters vary for every target position. The intrinsic parameters are evaluated by a direct linear transformation method that commonly incorporates sum-of-squares optimisation of the detected feature points of the artefact (Hartley and Zisserman 2003). Due to the high number of feature points that can be gathered and the unoccupied measurement volume to place targets, the pre-calibration is more accurate than other intrinsic computation methods carried out during the measurement of objects (Chiodini *et al* 2018)

The extrinsic parameters are measured during the measurement and are necessary for registration of 3D points to the measured object coordinate system. How the extrinsic parameters are measured varies from one LT system to another, and their accuracies significantly affect the overall accuracy of the LT system (Sims-Waterhouse *et al* 2019).

Commercial and in-house LT systems make use of existing metrological infrastructures, such as contact CMSs (Chekh *et al* 2019, Isheil *et al* 2011), articulated arm CMSs (Cuesta *et al* 2019) and laser trackers (Fernandez *et al* 2018), to determine the extrinsic parameters. Systems for measuring small-scale parts, typically using linear and rotary stages, need additional processes to determine the geometric parameters that define the kinematics of the motion systems (Isa and Lazoglu 2017).

Multi-camera systems can triangulate laser features without the need for the geometric properties of the laser. However, the camera-laser triangulation can provide additional information about a measured part. When triangulation of the laser is involved, the geometric parameters of the laser(s) used also need to be configured. The laser geometric parameters—the triangulation angle and laser baseline distance—are commonly determined by fitting of the measured laser points on lines or planes (Idrobo-Pizo *et al* 2019, Wu *et al* 2020).

3.5.4 Scanning path planning

Scanning path planning may be required for flexible and automated LT systems. Some LT systems, such as those with fixed 1D scanning motion, have a simple path that does not require planning. Robotic LT systems can be programmed to scan over a part using a desired LT sensor orientation along a desired path. The path planning can be based on some prior information about the part, such as a CAD model. Search algorithms have been used on triangular mesh models to plan the scanning path (Li *et al* 2019). LT sensor paths can be determined by taking into account the optimum focusing range of the sensor to improve accuracy. Scanning paths can incorporate the reflective properties of the part to provide better point density and accuracy. Guidelines on LT sensor orientation to minimise outlier formation on reflective surfaces are given elsewhere (Wang and Feng 2014).

3.5.5 Image pre-processing

Before features on an image are analysed to extract laser peak points and curves, the image usually undergoes a preparation stage. This stage normally includes filtration, smoothing and contrast adjustment.

It is common to use bandpass filters to allow only light of the specified laser wavelengths of the electromagnetic spectrum to reach the image sensor. Coloured images can be converted to greyscale from the standard coefficients for the construction of luminance (ITU 2017), given in

$$\boldsymbol{I} = 0.299\boldsymbol{I}_R + 0.587\boldsymbol{I}_G + 0.114\boldsymbol{I}_B, \tag{3.25}$$

where the values $\boldsymbol{I}_R$, $\boldsymbol{I}_G$ and $\boldsymbol{I}_B$ are the red, green and blue channels of light intensity in a colour image, respectively. Different coefficients can be used to obtain the optimum representation of the laser light against the image background. When needed, contrast can be adjusted to accentuate the laser irradiated regions in the image.

Image sensors have noise that affects images and measurement consistency. A popular method for mitigating the noise is by using smoothing filters. To smooth the effects of noise, a Gaussian kernel $\boldsymbol{h}$ is convolved with greyscale image $\boldsymbol{I}(i, j)$ in equation (3.25), thus

$$\boldsymbol{I}(i, j) * \boldsymbol{h} = \sum_{k=1}^{n_1}\sum_{l=1}^{n_2}\boldsymbol{h}(k, l)\boldsymbol{I}(i - k, j - l). \tag{3.26}$$

The $n_1 \times n_2$ sized filter $\boldsymbol{h}$ in equation (3.26) is expressed by

$$\boldsymbol{h}(k, l) = \frac{e^{\frac{(k^2+l^2)}{2\sigma^2}}}{\sum_{k=1}^{n_1}\sum_{l=1}^{n_2}e^{\frac{(k^2+l^2)}{2\sigma^2}}}. \tag{3.27}$$

The size $n_1 \times n_2$ and standard deviation σ should be determined according to the width and intensity of the detected laser stripe (Molleda *et al* 2013), and the greyscale image is convolved with h discretely. The effect of the filter on a thinned laser line in a sample image can be seen in figure 3.18. The resulting centreline of the image is smoother when the filter is applied and has less noise.

3.5.6 Laser feature extraction

In LT, the topographical information of a surface is contained in the distribution of the pixels in the images that encode the laser spots or stripes reflected from the surface. The distributions of the spots and the stripe sections across the width emulate the Gaussian beam intensity profile of the laser diode source. The task in LT is to determine the peak positions. Images with laser stripes (or spots) need to be processed and the intersection between the measured object and the laser central plane (or line) obtained. The external lighting should be as uniform as possible, and exposure should be adjusted for every change in the scanning condition.

Determination of the peak positions usually requires a series of image processing steps that begin with image segmentation. Segmentation separates the region of

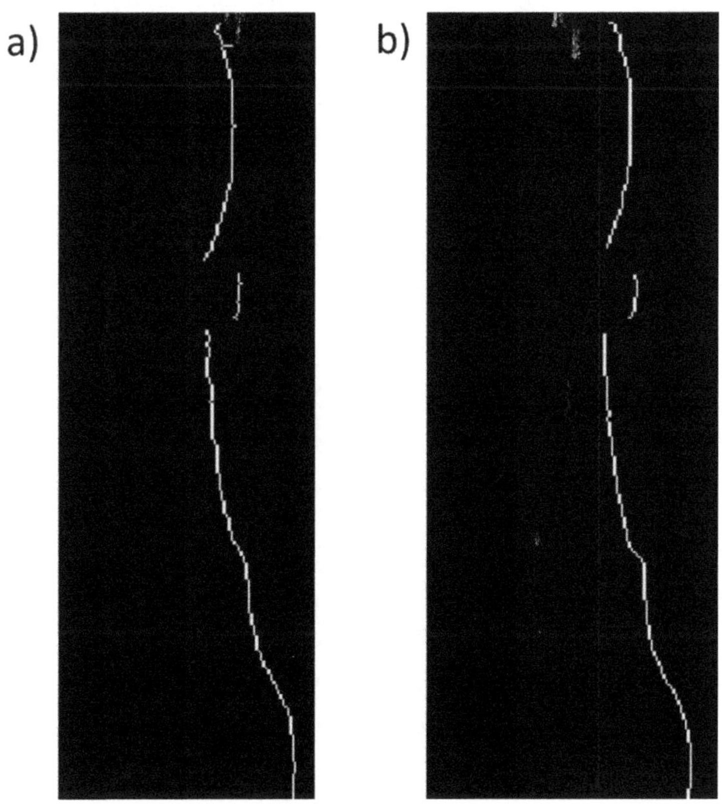

Figure 3.18. Comparison of (a) unfiltered and (b) filtered images, using a filter size of six pixels × six pixels and standard deviation σ = 3 pixels, after a thinning morphological operation.

interest (ROI) from the background. A global Otsu thresholding (Otsu 1979) determines the threshold that maximises the variance between the ROI and the background. Segmented ROIs may contain higher order reflections; hence, a selection algorithm can be used for isolation of the region of the primary laser reflection. Segmentation of a spot is much simpler than a stripe because the reflection of a spot has a more consistent shape than that from a reflected stripe.

In the neighbourhood-based segmentation shown in figure 3.19, thresholding of the intensity image is carried out locally within an appropriately sized window that trails along the laser stripe (Isa and Lazoglu 2017). The local thresholding is adaptive because, for regions where the angle of the surface normal is wide compared to the observation direction, the segmentation takes into account the relatively low intensity of observed laser light (intensity decreases as observation angle increases from Lambertian reflection). Hence, the use of a single global threshold limit can eliminate regions with high observation angles.

The adaptive segmentation process begins by an approximate segmentation that is carried out to isolate regions that exceed a certain threshold. The initial threshold is chosen to allow all relevant portions of the image to be included. For this purpose, the threshold value is chosen to be a fraction of the Otsu global threshold to obtain a binary image. A morphological filling of the generated binary image is performed after which identification of regions is carried out and the image's connected components are labelled accordingly. The mid-points of the paths are found row by row for each connected object in the image. Decisions on which region to use have to be made when more than one connected component is found on a row; properties of the components, such as area, are used to discriminate between the

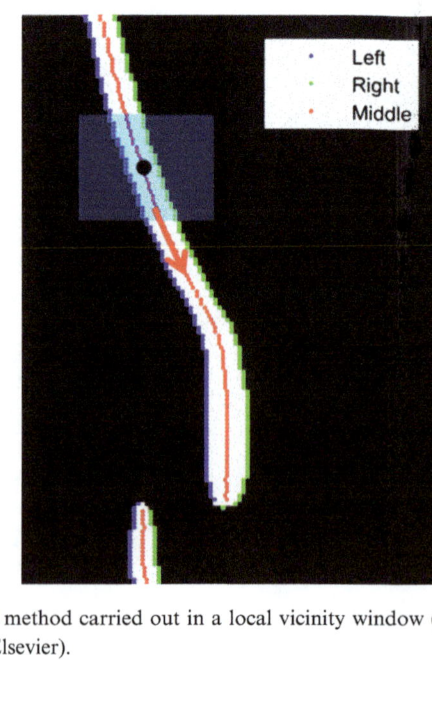

Figure 3.19. Local threshold method carried out in a local vicinity window (reprinted from Isa and Lazoglu 2017 with permission from Elsevier).

components. With the mid-points defined, thresholding limits are re-evaluated locally for each small window with the dimensions variably chosen according to stripe width.

The accurate boundary of a laser stripe can be defined using edge detection filters, such as the Canny edge detection filter or the neighbourhood-adaptive segmentation (Isa and Lazoglu 2017). Extraction methods, where each row (or column for a horizontal laser line) is analysed separately to extract the stripe, have been considered (Molleda *et al* 2013, Usamentiaga *et al* 2012). Sub-pixel approximation methods using interpolation of pixel location based on Gaussian and centre of gravity models were shown to have comparable performance (Usamentiaga *et al* 2012, Forest *et al* 2004).

The centre of gravity method is the most popular method used to find the detection of peak points from the reflection of laser light. The centre of gravity of a stripe is the intensity weighted centroid of pixel positions across the stripe. The Gaussian peak extraction method fits the row data into a normal distribution function and the peak point is found (Qi *et al* 2013). For further reference, various other stripe detection methods are suggested using, for example a Hessian matrix (Wu *et al* 2020) and Gaussian derivative (Colak *et al* 2018).

3.5.7 Refinement and postprocessing

After extraction of laser peaks from intensity distributions in an image, the LT models introduced in section 3.4 can be used to reconstruct 3D points. For multi-camera LT systems, further refinement of 3D points is possible through bundle adjustment (Hartley and Zisserman 2003). Bundle adjustment is a photogrammetric optimisation tool that minimises overall reprojection error by tuning the parameters of the cameras. For n 3D points $\boldsymbol{X}_{1,2,\ldots,n}$, the function $\boldsymbol{f}$ projects the points through the m cameras with intrinsic parameters $\boldsymbol{K}^{1,2,\ldots m}$ and poses $\boldsymbol{T}^{1,2,\ldots m}$ to minimise the sum

$$\min_{\boldsymbol{K}^{1,2,\ldots m},\boldsymbol{T}^{1,2,\ldots m}} \sum\nolimits_{j=1}^{m}\sum\nolimits_{i=1}^{n}(\boldsymbol{f}(\boldsymbol{X}_i, \boldsymbol{K}^j, \boldsymbol{T}^j) - \boldsymbol{U}_i^j)^2. \tag{3.28}$$

U_i^j represents the observed image point of X_i of camera j in equation (3.28). Efficient implementation of bundle adjustment can be carried out using a Ceres solver (Wilson *et al* 2017). Bundle adjustment is discussed further in (Luhmann *et al* 2011, Hartley and Zisserman 2003, Förstner and Wrobel 2016).

The quality of reconstructed 3D points is assessed using specific metrics: point density, completeness, noise and accuracy (Lartigue *et al* 2002). Points are commonly structured into a point cloud that is used to generate meshes and CAD models (see chapter 2). Acquired points can be organised into a mesh by generating neighbouring vertices from points (Woo *et al* 2002). Meshes could be further processed into any CAD format that can be used in solid modelling software packages. Reconstruction of an entire object requires registration and fusion of multiple point clouds or meshes. Figure 3.20 shows a sphere and a knight object that were reconstructed using the processes outlined in this section (Isa and Lazoglu 2017). For more on point clouds and meshes, see chapter 2.

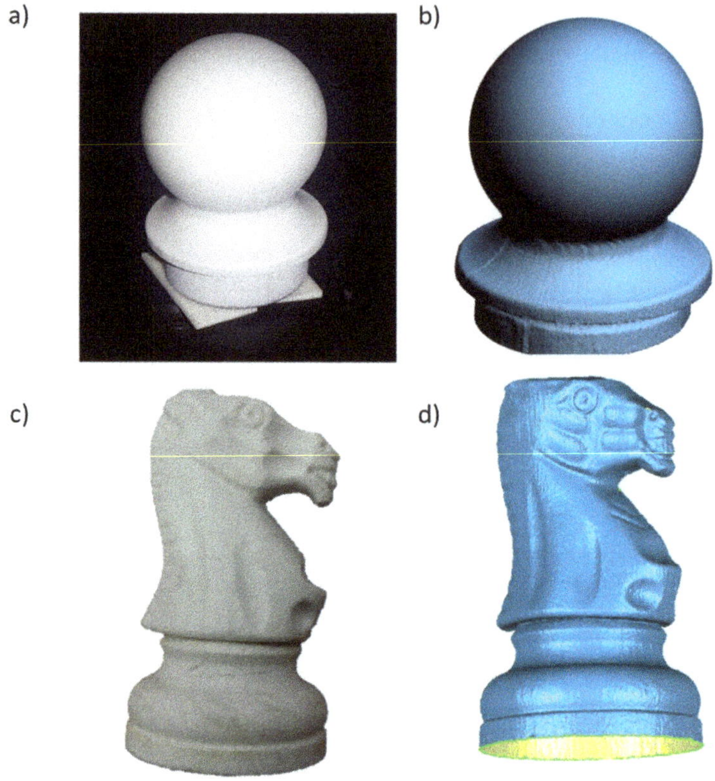

Figure 3.20. Reconstruction of a spherical artefact and a knight chess piece (reprinted from Isa and Lazoglu 2017 with permission from Elsevier).

3.6 Application of laser triangulation measurements

Applications of LT measurement results are considered in this section. LT measurement results can be applied at various stages of the reconstruction process; the applied result can be in the form of dense points, derived geometric dimensions or reconstructed CAD models. Obtaining dense measurement is often necessary for parts produced using emerging manufacturing methods because of the geometric complexity of the parts.

3.6.1 Application of laser triangulation in emerging manufacturing methods

Within the context of the fourth Industrial Revolution, where individualised production seeks to supplant mass production (Vaidya *et al* 2018), the fabrication of highly complex parts poses a challenge for quality verification. Due to emerging manufacturing methods, such as additive manufacturing (AM), that have gained a significant foothold, complex parts can be fabricated (Leach and Carmignato 2020). These parts are commonly designed for an optimised topology that improves the functionality of the part. Based on international measurement standards, the production of such complex parts still relies on contact-type quality inspection that can be slow and limited to measuring certain features. The measurement of complex topographies can be carried out using non-contact methods, such as LT, which are more suited for digitisation of surfaces because of the high speed and adaptability of the measurement methods. In addition, due to the lack of physical contact with a measured part, LT can be used to measure parts that are subjected to harsh working conditions for both a human operator and the measurement device. Fast inspection of parts at elevated temperatures (Ghiotti *et al* 2015) can be carried out to aid in making informed decisions in industrial automation (Fernandez *et al* 2019). LT systems can be used to measure surfaces with sharp, serrated features that are hazardous to a contact-type CMS. For instance, LT has been used for measurement of sharp effusion holes on the casing of the tubes for an aircraft combustion chamber (Lampa *et al* 2017). Remote LT systems provide a convenient means of inspection of chemically reactive and radioactive substances (Diggins *et al* 2015).

Manufacturing is headed towards distributed system architectures that incorporate multi-stage manufacturing requiring adaptable process and quality control (Xu *et al* 2018). LT systems present an opportunity for automation by verification and detection of manufacturing process steps. The necessary checks-and-balances, diagnosis and monitoring of processes can be carried out by fast autonomous LT measurements. Hence, by measuring distance, 2D profiles and 3D surfaces, important operational decisions can be made. It has been demonstrated that LT systems can be integrated with machine tools and measurements can be carried out on-machine (Savin *et al* 2018, Kou *et al* 2020). For instance, an LTS integrated into a laser metal deposition set-up has been used for monitoring the deposition height in an AM process (Donadello *et al* 2019). The monitored height can be used for real-time regulation of process parameters to reduce defects and improve production quality. Hence, the integration of LT systems into industrial facilities can improve in-process inspection and give feedback to improve efficiency.

While LT sensors can be integrated into existing industrial lines to accomplish a specific measurement task (So *et al* 2012), there are complete LT measurement solutions that are commercially available. Recent commercial LT systems allow manipulation of the LT sensor in six degrees of freedom. These products offer proprietary approaches for accurate registration of the LT sensor during high-speed measurement. Several commercial LT systems implement photogrammetric tracking of hand-held or robot manipulated LT sensor (for example, Creaform n.d.). Research on registration of hand-held LT sensors without tracking has also been explored (Arold *et al* 2009, Huber *et al* 2010, Ettl *et al* 2012); however, the accuracy of the registration algorithms does not match that of tracking methods. The major impediment to industrial adoption of LT systems is the inadequate standardisation both in terms of physical metrological artefacts and universal algorithmic procedures (Novak 2014). There are internal standards used by suppliers and researchers but an accepted specification standard is still lacking (Phillips *et al* 2009, Carmignato *et al* 2020). There is limited research to define calibration standards similar to contact-type measurement standards (Genta *et al* 2016, Shen *et al* 2020), however, these approaches may not fully cover the capacity of optical CMS. Existing specification standards for performance verification of optical CMS measurements are discussed in chapter 8 of this book.

3.6.2 Application for geometric inspection

Geometric inspection of manufactured products can be carried out using the results obtained by LT systems. Point measuring LT systems provide a portable alternative to common tabletop contact CMSs for fast measurement of 1D features. From reconstructed points, 1D features such as height, thickness, gaps and flushes can be measured. Profile measuring LT systems can also be used for measuring 2D features such as circles, lines and angles. These profile measurements can be incorporated in extrusion, rolling and rotating parts to acquire cross-sectional profiles during industrial processes. For complete 3D surfaces, LT systems can generate dense 3D point clouds of surfaces which can be used to derive geometric dimensions and tolerances.

Point clouds from LT systems are used for defect detection on produced parts. LT triangulation results are used to zone defect areas on parts which, when combined with image processing, can be used for classification of defects (Wang *et al* 2020, Jovančević *et al* 2017). A general inspection approach is to compare the distances of the points from the design model of the manufactured part. The point cloud has to be aligned to the model by determining the position and orientation that results in the minimum sum of distances. The alignment process is commonly carried out by the iterative closest point (ICP) algorithm (see chapter 2). Figure 3.21(a) shows a 50 mm by 50 mm Ti-6Al-4V part produced by powder bed fusion and measured by LT. The laser stripes, shown in figure 3.21(b), were acquired by rotating the part and

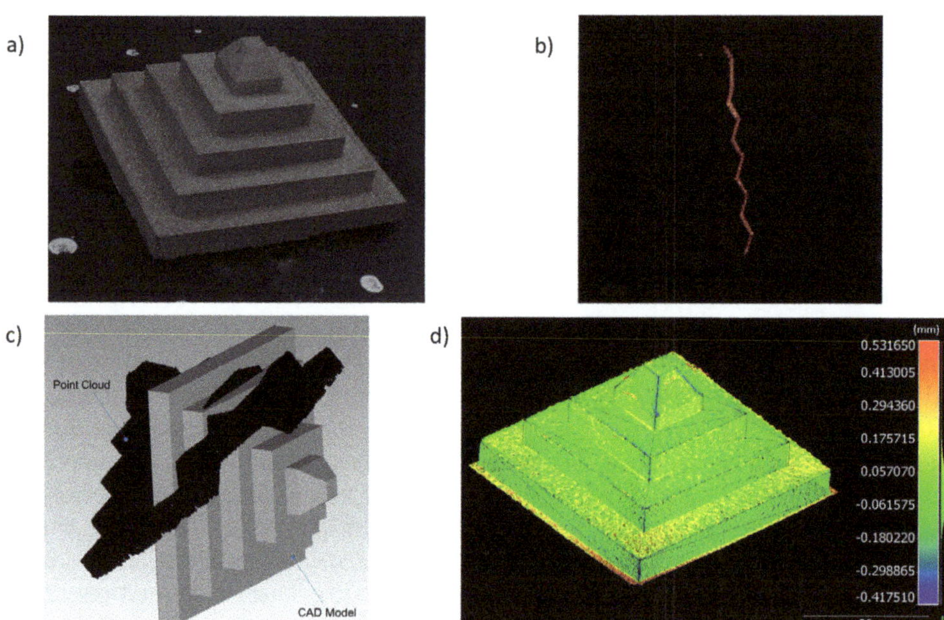

Figure 3.21. 3D measurement of an additive manufactured artefact in (a) using laser stripes given in (b) to generate point cloud in (c). The point cloud is registered to the CAD model and the point-to-model deviations are given in (d).

the point cloud, given in figure 3.21(c), was generated. After alignment of the point cloud to the CAD model, the distribution of the point-to-model deviations is given in figure 3.21(d).

In addition to the analysis of point-to-model distances, regions in a point cloud can be used to analyse the form error of a part. A comparison of the flatness measurement from an LT system with contact CMS is shown in figure 3.22 (Brosed *et al* 2011). The 125 mm × 96 mm flange face of a machined part is measured using a robot actuated LT sensor (Brosed *et al* 2011). Other types of geometric dimensioning and tolerancing, such as the positions and circularity of holes, can also be analysed from the point cloud.

3.6.3 Application of 3D reconstructed models

The process of reconstruction of a virtual model is particularly important for two classes of objects. The reconstructed virtual models are commonly in the form of meshes or solid models, in computer file formats such as STL and STEP, respectively.

The first class consists of objects that do not have any computer models. Such objects do not undergo any systematic product development process. They can be naturally created entities—such as a human face, a tooth or plant leaves—or obsolete objects that predate modern engineering. Whenever these parts need to be reused or engineered, their shape needs to be reconstructed. For instance, when designing a prosthetic for a human limb, it is necessary to reconstruct the stump and match the interface region of the prosthetic with the limb stump (Ryniewicz *et al* 2017). A befitting design of the prosthetic can be customised and manufactured based on the reconstructed models. Consumer-specific products, such shoe soles, have also been manufactured using reconstructed models of feet (Wang 2010). In addition, attempts at the reconstruction of human faces from images to analyse identity or emotion are more examples of the application of natural models in engineering settings. Applications seeking to interface natural objects with the digital world also rely on reconstructed CAD models, such as computer generated

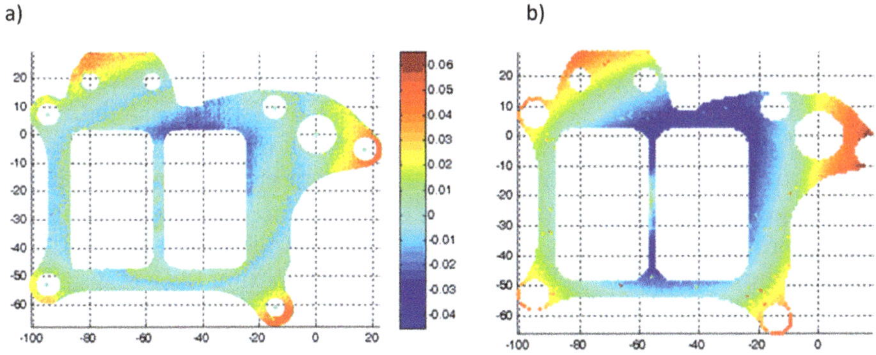

Figure 3.22. Deviation of (a) LT measured points and (b) contact CMS measured points from a theoretical plane (reprinted from Brosed *et al* 2011 with permission from MDPI).

imagery (popularly known as CGI) in the entertainment industries, reverse engineering in early product design, individual-specific digital model integration in medicine and augmented reality (Tavares *et al* 2019). The first class of objects also includes natural geographical objects, where fields such as landscape mapping, simultaneous localisation and mapping (SLAM) and urban reconstruction can benefit from geographical computer models (Yang *et al* 2019). There are also obsolete components and objects whose production predates the modern computerised design. Applications in archaeology, art and historic monument preservation benefit from virtual models obtained through LT (Raimundo *et al* 2018).

The second class of objects are parts with known design CAD model or shape. 3D reconstructions of these objects are commonly used for inspection and quality control. In high-precision and microscale manufacturing, LT may not be useful for direct quality control of fabricated surfaces, but it can be used for measurement of assembled components. Tolerances on assembled components can be wide enough to make LT sensors applicable for quality control and inspection at several stages of manufacturing (Minnetti *et al* 2020). In addition to end-product inspection, there are objects that undergo significant shape changes during application. These include objects that are subjected to high continuous stresses or high accidental loads. Manufactured parts, dies and moulds used in forming processes, such as forging and casting, can be reconstructed using LT. Deformation of parts during a forming process can be monitored using integrated LT systems to aid in controlling shrinkage (Ding *et al* 2016). The non-contact nature of LT enables the coordinate measurement of non-solid products, such as volume measurement of bulk powder materials (Min *et al* 2020). In the transport industries, the deformation of tyres, railway and train tunnels are monitored using LT sensors (Farahani *et al* 2019). The portability and adaptability of LT systems opens up avenues for *in situ* integration for quality control and monitoring. Defective products, that either reach the end of their product life cycle or are impaired by accident, can also benefit from LT reconstruction (Yeo *et al* 2017) as part of the remanufacturing process. The reconstruction of faulty products plays an important role in remanufacturing which enhances the sustainability in product life cycle management.

3.7 Conclusions

The future manufacturing ecosystem is envisioned as cyber–physical systems implementing new technologies, such as the Internet of Things, big data, cloud computing and artificial intelligence, that enhance the information flow between machines and virtual systems. Within this industrial framework, quality inspection using current contact CMSs is inadequate in terms of speed and adaptability for individualised products. Therefore, non-contact optical methods, such as LT, are expected to supplant the present tabletop contact CMSs. Within the last decade, LT systems have been advanced by implementing more accurate and flexible positioning systems, thereby increasing their application in industry.

The capacity to digitise complex geometries with high speed is a futuristic attribute of LT in the manufacturing industry. With improved knowledge on handling large point clouds using new technologies in big data and artificial intelligence, the digitisation of manufactured parts can become an integral part of smart manufacturing.

This chapter sums up the relevant work on the dependence of LT measurements on surface properties, the application of the triangulation principle in various LT systems and how the measurement systems are used in different applications.

References

Amir Y M and Thörnberg B 2017 High precision laser scanning of metallic surfaces *Int. J. Opt.* **2017** 4134205

Arold O, Yang Z, Ettl S and Häusler G 2009 A new registration method to robustly align a series of sparse 3D data *DGaO Proc. (Erlangen-Nürnberg, Germany)* pp 20–1

Beckmann P and Spizzichino A 1963 *The Scattering of Electromagnetic Waves from Rough Surfaces* (New York: Pergamon)

Bertani D, Cetica M, Ciliberto S and Francini F 1984 High-resolution light spot localization with photodiode arrays *Rev. Sci. Instrum.* **55** 1270–2

Black J T and Kohser R A 2011 *Materials and Processes in Manufacturing* (New York: Wiley)

Brosed F J, Aguilar J J, Guillomïa D and Santolaria J 2011 3D geometrical inspection of complex geometry parts using a novel laser triangulation sensor and a robot *Sensors* **11** 90–110

Cajal C, Santolaria J, Samper D and Garrido A 2015 Simulation of laser triangulation sensors scanning for design and evaluation purposes *Int. J. Simul. Model.* **14** 250–64

Carmignato S, De Chiffre L, Bosse H, Leach R K, Balsamo A and Estler W T 2020 Dimensional artefacts to achieve metrological traceability in advanced manufacturing *CIRP Ann.* **69** 693–716

Chekh B A, Kortaberria G and Gonzalo O 2019 Extrinsic calibration and kinematic modelling of a laser line triangulation sensor integrated in an intelligent fixture with 3 degrees of freedom *Precis. Eng.* **60** 235–45

Chiodini S, Marco P, Giubilato R, Salviolli F, Berrera M, Franceschetti P and Debei S 2018 Camera rig extrinsic calibration using a motion capture system *2018 5th IEEE Int. Workshop on Metrology for AeroSpace (Rome)* pp 590–5

Colak S, Fresse V, Alata O and Gautrais T 2018 Comparative study of laser stripe detection algorithms for embedded real-time suitability in an industrial quality control context *J. Phys.* **1074** 012173

Costa M F M 2012 Optical triangulation-based microtopographic inspection of surfaces *Sensors* **12** 4399–420

Craggs G, Meuret Y, Danckaert J and Verschaffelt G 2012 Low speckle line generation using a semiconductor laser source *Proc. SPIE* **8433** 84330M

Creaform n.d. *Creaform MetraScan 3D* https://creaform3d.com/en/optical-3d-scanner-metrascan

Cuesta E, Alvarez B J, Martinez-Pellitero S, Barreiro J and Patiño H 2019 Metrological evaluation of laser scanner integrated with measuring arm using optical feature-based gauge *Opt. Lasers Eng.* **121** 120–32

Curless B L 1997 *New Methods for Surface Reconstruction from Range Images* (Berkeley, CA: Stanford University)

Diggins Z J, Mahadevan N, Herbison D, Karsai G, Barth E, Reed R A, Schrimpf R D, Weller R A, Alles M L and Witulski A 2015 Range-finding sensor degradation in gamma radiation environments *IEEE Sens. J.* **15** 1864–71

Ding D, Zhao Z, Zhang X, Fu Y and Xu J 2020 Evaluation and compensation of laser-based on-machine measurement for inclined and curved profiles *Measurement* **151** 107236

Ding Y, Zhang X and Kovacevic R 2016 A laser-based machine vision measurement system for laser forming *Measurement* **82** 345–54

Donadello S, Motta M, Demir A G and Previtali B 2019 Monitoring of laser metal deposition height by means of coaxial laser triangulation *Opt. Lasers Eng.* **112** 136–44

Dong Z, Sun X, Liu W and Yang H 2018 Measurement of free-form curved surfaces using laser triangulation *Sensors* **18** 3527

Donges A and Noll R 2015 Laser triangulation ed A Donges and R Noll *Laser Measurement Technology* (Berlin: Springer)

Dorsch R G, Häusler G and Herrmann J M 1994 Laser triangulation: fundamental uncertainty in distance measurement *Appl. Opt.* **33** 1306–14

Drap P and Lefèvre J 2016 An exact formula for calculating inverse radial lens distortions *Sensors* **16** 807

Drouin M-A and Beraldin J-A 2012 Active 3D imaging systems ed P Nick, L Yonghuai and P Bunting *3D Imaging Analysis and Applications* (London: Springer)

Du S and Xi L 2019 *High Definition Metrology Based Surface Quality Control and Applications* (Berlin: Springer)

Ettl S, Arold O, Yang Z and Häusler G 2012 Flying triangulation—a motion-robust optical 3D sensor for the real-time shape acquisition of complex objects *Appl. Opt.* **51** 281–9

Farahani B V, Barros F, Sousa P J, Cacciari P P, Tavares P J, Futai M M and Moreira P 2019 A coupled 3D laser scanning and digital image correlation system for geometry acquisition and deformation monitoring of a railway tunnel *Tunn. Undergr. SP Technol.* **91** 102995

Fernandez A, Souto M A and Guerra L 2019 Automatic steel bar counting in production line based on laser triangulation *IECON 2019–45th Annual Conf. IEEE Industrial Electronics Society (Lisbon)* pp 80–5

Fernandez S R, Olabi A and Gibaru O 2018 On-line accurate 3D positioning solution for robotic large-scale assembly using a vision system and a 6DoF tracking unit *Proc. IEEE 3rd Advanced Information Technology, Electronic and Automation Control Conf. (Chongqing, China)* pp 682–8

Forest J, Salvi J, Cabruja E and Pous C 2004 Laser stripe peak detector for 3D scanners. A FIR filter approach *Proc. 17th Int. Conf. on Pattern Recognition (Cambridge)* pp 646–9

Förstner W and Wrobel B P 2016 *Photogrammetric Computer Vision: Statistics, Geometry, Orientation and Reconstruction* (Cham: Springer International)

Gao F, Lin H, Chen K, Chen X and He S 2018 Light-sheet based two-dimensional Scheimpflug lidar system for profile measurements *Opt. Express* **26** 27179

Genta G, Minetola P and Barbato G 2016 Calibration procedure for a laser triangulation scanner with uncertainty evaluation *Opt. Lasers Eng.* **86** 11–9

Ghiotti A, Schöch A, Salvadori A, Carmignato S and Savio E 2015 Enhancing the accuracy of high-speed laser triangulation measurement of freeform parts at elevated temperature *CIRP Ann.* **64** 499–502

Giganto S, Martínez-Pellitero S, Cuesta E, Meana V M and Barreiro J 2020 Analysis of modern optical inspection systems for parts manufactured by selective laser melting *Sensors* **20** 3202

Goodman W 1975 *Laser Speckle and Related Phenomena* (Berlin: Springer)

Gruber F, Wollmann P, Grählert W and Kaskel S 2018 Hyperspectral imaging using laser excitation for fast Raman and fluorescence hyperspectral imaging for sorting and quality control applications *J. Imaging* **4** 110

Hartley R and Zisserman A 2003 *Multiple View Geometry in Computer Vision* (Cambridge: Cambridge University Press)

Häusler G, Ettl P, Schenk M, Bohn G and Laszlo I 1999 Limits of optical range sensors and how to exploit them *International Trends in Optics and Photonics* ed T Asakura (Berlin: Springer)

Hausler G and Ettl S 2011 Limitations of optical 3D sensors *Optical Measurement of Surface Topography* ed R K Leach (Berlin: Springer)

Huber F, Arold O, Willomitzer F, Ettl S and Häusler G 2010 3D body scanning with 'flying triangulation' *DGaO Proc. (Erlangen-Nürnberg, Germany)* pp 1–2

Idrobo-Pizo G A, Motta J M S T and Sampaio R C 2019 A calibration method for a laser triangulation scanner mounted on a robot arm for surface mapping *Sensors* **19** 1783

Isa M A 2018 *Multi-Axis Additive Manufacturing and 3D Scanning of Freeform Models* (Istanbul: Koc university)

Isa M A and Lazoglu I 2017 Design and analysis of a 3D laser scanner *Measurement* **111** 122–33

Isheil A, Gonnet J-P, Joannic D and Fontaine J-F 2011 Systematic error correction of a 3D laser scanning measurement device *Opt. Lasers Eng.* **49** 16–24

ITU 2017 *Studio Encoding Parameters of Digital Television for Standard 4:3 and Wide Screen 16:9 Aspect Rations ITU-R BT.601-7* (Geneva: ITU)

Ji Z and Leu M C 1989 Design of optical triangulation devices *Opt. Laser Technol.* **21** 339–41

Johannes S, Csencsics E and Georg S 2018 Optical scanning of laser line sensors for 3D imaging *Appl. Opt.* **57** 5242–8

Jovančević I, Pham H H, Orteu J J, Gilblas R, Harvent J, Maurice X and Brèthes L 2017 3D point cloud analysis for detection and characterization of defects on airplane exterior surface *J. Nondestruct. Eval.* **36** 74–91

Kou M, Wang G, Li W and Mao J 2020 Calibration of the laser displacement sensor and integration of on-site scanned points *Meas. Sci. Technol.* **31** 125104

Lampa P, Mrzygłód M and Reiner J 2017 Triangulation methods for effusion holes measurements in combustion chambers of aircraft engines *Mechanik* **90** 1164–8

Lartigue C, Contri A and Bourdet P 2002 Digitised point quality in relation with point exploitation *Measurement* **32** 193–203

Latimer W 2015 Understanding laser-based 3D triangulation methods *Vis. Syst. Des.* **20** 31–5

Leach R K 2011 *Optical Measurement of Surface Topography* (Berlin: Springer)

Leach R K and Carmignato S 2020 *Precision Metal Additive Manufacturing* (Boca Raton, FL: CRC Press)

Li B, Li F, Liu H, Cai H, Mao X and Peng F 2014 A measurement strategy and an error-compensation model for the on-machine laser measurement of large-scale free-form surfaces *Meas. Sci. Technol.* **25** 015204

Li L, Xu D, Niu L, Lan Y and Xiong X 2019 A path planning method for a surface inspection system based on two-dimensional laser profile scanner *Int. J. Adv. Robot. Syst.* **16** 1–13

Li S, Yang Y, Jia X and Chen M 2016 The impact and compensation of tilt factors upon the surface measurement error *Optik* **127** 7367–73

Li Y, Kästner M and Reithmeier E 2018 Triangulation-based edge measurement using polyview optics *Opt. Lasers Eng.* **103** 71–6

Lim J S and Nawab H 1981 Techniques for speckle noise removal *Opt. Eng.* **20** 472–80

Lin H, Wang H, Zhu X, Zhu G and Qi L 2016 Design of homogeneous laser-line-beam generators *Opt. Eng.* **55** 095106

Luhmann T, Robson S, Kyle S and Harley I 2011 *Close Range Photogrammetry: Principles, Techniques and Applications* (Dunbeath, UK: Whittles Publishing)

Martínez S, Cuesta E, Barreiro J and Álvarez B 2010 Methodology for comparison of laser digitizing versus contact systems in dimensional control *Opt. Lasers Eng.* **48** 1238–46

Matharu R S, Sadler W, Gashi B V and Toman T 2019 Investigation in optimisation of accuracy with non-contact systems by influencing variable processes *19th Int. Congress of Metrology (Paris)* 09004

Min F, Lou A and Wei Q 2020 Design and experiment of dynamic measurement method for bulk material of large volume belt conveyor based on laser triangulation method *IOP Conf. Ser. Mater. Sci. Eng.* **735** 012029

Minnetti E, Chiariotti P, Paone N, Garcia G, Vicente H, Violini L and Castellini P 2020 A smartphone integrated hand-held gap and flush measurement system for in line quality control of car body assembly *Sensors* **20** 1–17

Molleda J, Usamentiaga R, García D F, Bulnes F G, Espina A, Dieye B and Smith L N 2013 An improved 3D imaging system for dimensional quality inspection of rolled products in the metal industry *Comput. Ind.* **64** 1186–200

Mueller T, Poesch A and Reithmeier E 2015 Measurement uncertainty of microscopic laser triangulation on technical surfaces *Microsc. Microanal.* **21** 1443–54

Nayar S K, Ikeuchi K and Kanade T 1991 Surface reflection: physical and geometrical perpectives *IEEE Trans. Pattern Anal. Mach. Intell* **13** 611–34

Novak E 2014 Advanced defect and metrology solutions *Proc. SPIE* **9110** 91100G

Okada T 1982 Optical distance sensor for robots *Int. J. Robot. Res.* **1** 3–14

Otsu N 1979 A threshold selection method from gray-level histograms *IEEE Trans. Syst. Man. Cybern* **9** 62–6

Pavlicek P and Hybl O 2008 White-light interferometry on rough surfaces—measurement uncertainty caused by surface roughness *Appl. Opt.* **47** 2941–9

Pears N E 1994 Optical triangulation range sensors *Optical Triangulation Range Sensors for Vehicle Manoeuvres* ed S Probert (Oxford: World Scientific Publishing)

Peiravi A and Taabbodi B 2010 A reliable 3D laser triangulation-based scanner with a new simple but accurate procedure for finding scanner parameters *J. Am. Sci.* **6** 80–5

Pereira J R M, de Lima e Silva Penz I and da Silva F P 2019 Effects of different coating materials on three-dimensional optical scanning accuracy *Adv. Mech. Eng.* **11** 1–6

Peterson J P and Peterson R B 2006 Laser triangulation for liquid film thickness measurements through multiple interfaces *Appl. Opt.* **45** 4916–26

Petrov M, Talapov A, Robertson T, Lebedev A, Zhilyaev A and Polonskiy L 1998 Optical 3D digitizers: bringing life to the virtual world *IEEE Comput. Graph. Appl.* **18** 28–37

Phillips S, Krystek M, Shakarji C and Summerhays K 2009 Dimensional measurement traceability of 3D imaging data *Proc. SPIE* 7239 72390E

Qi L, Zhang Y, Zhang X, Wang S and Xie F 2013 Statistical behavior analysis and precision optimization for the laser stripe center detector based on Steger's algorithm *Opt. Express* **21** 13442–9

Raghavendra K, Manjaiah M and Balashanmugam N 2020 4D printing *Materials Forming, Machining and Post Processing* ed K Gupta (Cham: Springer International)

Raimundo P O, Apaza-Agüero K and Apolinário A L Jr. 2018 Low-cost 3D reconstruction of cultural heritage artifacts *Rev. Bras. Comput. Apl.* **10** 66–75

Reiner J and Stankiewicz M 2011 Evaluation of the predictive segmentation algorithm for the laser triangulation method *Metrol. Meas. Syst* **18** 667–8

Rioux M 1984 Laser range finder based on synchronize scanners *Appl. Opt.* **23** 3837–44
Ryniewicz A M, Ryniewicz A, Bojko Ł, Gołębiowska W, Cichoński M and Madej T 2017 The use of laser scanning in the procedures replacing lower limbs with prosthesis *Measurement* **112** 9–15
Savin V N, Stepanov V A and Shadrin M V 2018 High-speed multisensor method of measurement, control and 3D analysis of complex object shapes in production environment *Opt. Mem. Neural Netw.* **27** 40–5
Sawatari T 1976 Real-time noncontacting distance measurement using optical triangulation *Appl. Opt.* **15** 2821
Schlarp J, Csencsics E and Schitter G 2020 Optically scanned laser line sensor *IEEE Int. Instrumentation and Measurement Technology Conf. (Dubrovnik)* pp 1–6
Schöch A and Savio E 2019 High-speed measurement of complex shaped parts by laser triangulation for in-line inspection *Metrology-Precision* ed W Gao (Singapore: Springer)
Schwarte R, Heinol H, Buxbaum B, Ringbeck T, Xu Z and Hartmann K 1999 Principles of three-dimensional imaging techniques *Handbook of Computer Vision and Applications Sensors and Imaging* ed B Jahne, H Haußecker and P Geißler (London: Academic Press)
Schwenke H, Neuschaefer-Rube U, Pfeifer T and Kunzmann H 2002 Optical methods for dimensional metrology in production engineering *CIRP Ann.* **51** 685–99
Shen Y, Zhang X, Wang Z, Wang J and Zhu L 2020 A robust and efficient calibration method for spot laser probe on CMM *Measurement* **154** 107523
Sims-Waterhouse D, Isa M A, Piano S and Leach R K 2019 Uncertainty model for a traceable stereo-photogrammetry system *Precis. Eng.* **63** 1–9
Smolka F M and Caudell T P 1978 Surface profile measurement and angular deflection monitoring using a scanning laser beam: a noncontact method *Appl. Opt.* **17** 3284–9
So E W Y, Michieletto S and Menegatti E 2012 Calibration of a dual-laser triangulation system for assembly line completeness inspection *IEEE Int. Symp. on Robotic and Sensors Environments (Magdeburg)* pp 138–43
Song Z 2013 *Handbook of 3D Machine Vision* (Boca Raton, FL: CRC Press)
Sousa G B, Olabi A, Palos J and Gibaru O 2017 3D metrology using a collaborative robot with a laser triangulation sensor *Procedia Manuf.* **11** 132–40
Stavroulakis P I and Leach R K 2016 Invited review article: review of post-process optical form metrology for industrial-grade metal additive manufactured parts *Rev. Sci. Instrum* **87** 041101
Sun B and Li B 2016 Laser displacement sensor in the application of aero-engine blade measurement *IEEE Sens. J.* **16** 1377–84
Sun B, Zhu J, Yang L, Guo Y and Lin J 2017 Stereo line-scan sensor calibration for 3D shape measurement *Appl. Opt.* **56** 7905–14
Tavares P, Costa C M, Rocha L, Malaca P, Costa P, Moreira A P, Sousa A and Veiga G 2019 Collaborative welding system using BIM for robotic reprogramming and spatial augmented reality *Autom. Constr.* **106** 102825
Torrance K E and Sparrow E M 1966 Off-specular peaks in the directional distribution of reflected thermal radiation *J. Heat Transfer* **88** 223–30
Tu D, Jin P and Zhang X 2019 Geometrical model of laser triangulation system based on synchronized scanners *Math. Probl. Eng.* **2019** 3503192
Usamentiaga R, Molleda J and García D F 2012 Fast and robust laser stripe extraction for 3D reconstruction in industrial environments *Mach. Vis. Appl* **23** 179–96
Vaidya S, Ambad P and Bhosle S 2018 Industry 4.0—a glimpse *Procedia Manuf.* **20** 233–8

Vukasinovic N and Duhovnik J 2019 Optical 3D geometry measurements based on laser triangulation *Advanced CAD Modeling* (Cham: Springer)

Wang C S 2010 An analysis and evaluation of fitness for shoe lasts and human feet *Comput. Ind.* **61** 532–40

Wang C Y, Tan Q C and Guo R H 2014 Design and optimization of a linear laser beam *Lasers Eng.* **27** 373–81

Wang W, Cai Y, Wang H P, Carlson B E and Poss M 2020 Quality inspection scheme for automotive laser braze joints *Int. J. Adv. Manuf. Technol.* **106** 1553–66

Wang Y and Feng H-Y 2014 Modeling outlier formation in scanning reflective surfaces using a laser stripe scanner *Measurement* **57** 108–21

Wilson A, Ben-Tal G, Heather J, Oliver R and Valkenburg R 2017 Calibrating cameras in an industrial produce inspection system *Comput. Electron. Agric.* **140** 386–96

Woo H, Kang E, Wang S and Lee K H 2002 A new segmentation method for point cloud data *Int. J. Mach. Tools Manuf.* **42** 167–78

Wu X, Tang N, Liu B and Long Z 2020 A novel high precise laser 3D profile scanning method with flexible calibration *Opt. Lasers Eng.* **132** 105938

Xu L D, Xu E L and Li L 2018 Industry 4.0: state of the art and future trends *Int. J. Prod. Res.* **56** 2941–62

Yang M, Yang E, Zante R C, Post M and Liu X 2019 Collaborative mobile industrial manipulator: a review of system architecture and applications *Proc. 25th Int. Conf. on Automation & Computing (Lancaster)* pp 1–6

Yao Z, Xie J, Tian Y and Huang Q 2019 Using Hampel identifier to eliminate profile-isolated outliers in laser vision measurement *J. Sensors* **2019** 3823691

Yeo N C Y, Pepin H and Yang S S 2017 Revolutionizing technology adoption for the remanufacturing industry *Proc. CIRP* **61** 17–21

Zhang H, Ren Y, Liu C and Zhu J 2014 Flying spot laser triangulation scanner using lateral synchronization for surface profile precision measurement *Appl. Opt.* **53** 4405–12

Zhang Y, Liu W, Li X, Yang F, Gao P and Jia Z 2015 Accuracy improvement in laser stripe extraction for large-scale triangulation scanning measurement system *Opt. Eng.* **54** 105108

Zhang Z 2000 A flexible new technique for camera calibration *IEEE Trans. Pattern Anal. Mach. Intell* **22** 1330–4

Zhang Z and Li C 2015 Defect inspection for curved surface with highly specular reflection ed Z Liu, H Ukida, P Ramuhalli and K Niel *Integrated Imaging and Vision Techniques for Industrial Inspection: Advances and Applications* (London: Springer)

Zhou L, Waheed A and Cai J 1998 Correction technique to compensate the form error in 3D profilometry *Measurement* **23** 117–23

Zuo H and He S 2018 Double stage FPCB scanning micromirror for laser line generator *Mechatronics* **51** 75–84

Zuo H, Nia F H and He S 2017 SOIMUMPs micromirror scanner and its application in laser line generator *J. Micro/Nanolithogr. MEMS MOEMS* **16** 015501

IOP Publishing

Advances in Optical Form and Coordinate Metrology

Richard Leach

Chapter 4

Close-range photogrammetry

Danny Sims-Waterhouse, Richard Leach and Samanta Piano

Over recent decades, photogrammetry has become an important technology for form and coordinate metrology. The emergence of high-resolution digital cameras and high-speed computers has enabled the application of photogrammetry to become more prominent. In this chapter, the key developments in the application of close-range photogrammetry to form and coordinate metrology are discussed. The recent advancements in photogrammetry focus on the improvement of the calibration accuracy, optimising the reconstruction speed and furthering the understanding of measurement uncertainty in the photogrammetry process.

4.1 Introduction

Photogrammetry is defined by Luhmann *et al* as the science of measurement through the use of photographs (Luhmann and Robson 2011). The main principle of photogrammetry for form measurement is the triangulation of corresponding points using two or more images. Correspondences are determined through a variety of methods, including physical targets or the detection of image features. Stereo photogrammetry is the most common application of photogrammetry, in which the measurement process is applied using only two images. The simplicity of photogrammetry has meant that it has been used as a method of measurement since photography was invented in the 1830s (Luhmann and Robson 2011). For the first fifty years, photogrammetry was mainly used by architects as a fast and easy method of measuring the shape of buildings. The nature of the photographic techniques meant that building facade dimensions could be measured without the risk of typically dangerous manual methods in which workers would be exposed to hazards. It was not until the late 1800s that the scope of photogrammetry became widespread. The development of stereo comparators allowed topographic measurements to be made and became invaluable in the mapping of inaccessible mountainous regions (Luhmann and Robson 2011).

doi:10.1088/978-0-7503-2524-0ch4

It was not until the rise of digital computers that close-range photogrammetry could truly be considered for accurate close-range measurements. In this case, close-range is defined to be any measurements under 300 m in size (Luhmann and Robson 2011). The potential of close-range photogrammetry was now only limited by the reconstruction algorithms and the imaging process used. As a result, research into reconstruction and calibration algorithms flourished, giving rise to the modern methods seen today. The recent advancements in digital photography have also been significant.

4.1.1 Modern photogrammetry

Although millimetre-scale measurements are feasible with photogrammetry, modern applications are still predominately applied on much larger scales. The non-invasive nature of the method allows for the measurement of objects or surfaces that cannot be physically accessed or are easily damaged. Consequently, photogrammetry is widely used in archaeology, architecture, the automotive and aerospace industries, medicine, and crime scene and accident reconstruction. In archaeology, photogrammetry means that the 3D structure of delicate objects and buildings can be measured without physical contact (Remondino *et al* 2008). This means the historical significance of such objects can be determined without the potential to destroy or damage them. During the investigation of an accident or crime, there may be a point at which the only remaining evidence is photographic. Photogrammetry allows a scene to be reconstructed in 3D from a series of images. Applications of photogrammetry to accident scene modelling date back to as early as 1933 and are still used today (Fraser *et al* 2008, Luhmann and Robson 2011). The non-invasive nature of photogrammetry makes it well suited to medical applications. By taking a 3D scan of the human body prior to surgery, photogrammetry can be used to tailor reconstructive surgery for a patient (Gu *et al* 2015). Augmented reality during surgery can also be implemented to aid surgeons in the accurate removal of tumours. The 3D position of the tumour can be determined and separated from healthy tissue allowing complete removal (Bergeles *et al* 2014).

Most relevant to metrological applications are the uses of photogrammetry in automotive and aerospace engineering. Stereo photogrammetry and other optical techniques are widely used as means of quality control during machining and forming processes (Möller *et al* 2016, Wang *et al* 2018). By scanning a component through various techniques, a 3D model can be created from the resulting point cloud. The scan can then be compared against the original CAD model to ensure that the required specifications have been met (see chapter 2). Being able to produce precision manufactured components and highly accurate alignments is critical in modern manufacturing engineering (Luhmann 2010).

4.1.2 Camera projection theory

The aim of camera projection theory is to model how 3D space is projected onto a two-dimensional (2D) image. Stereo photogrammetry can then, using a camera model, take 2D images and project them back in 3D space to deduce some information about the original object. However, the direct transfer from 3D to 2D infers some loss of information. This loss of information is exemplified by the inability of a camera to capture depth information. A point on an image will correspond to any point along a line in space, therefore, when taking an image, the distance along this line is not measured. An object will also appear smaller or larger depending on how far away it is from the camera. An ambiguity in the depth information captured in an image will, therefore, produce a loss of scale in the lateral directions. It is possible to use corresponding pixels in two or more images to triangulate that point in 3D space based on their projections. This triangulation method forms the main principle of reconstruction in stereo photogrammetry to determine the projection of a pixel. Camera models provide the mathematical basis for the geometric projections and hence must be accurately characterised for accurate 3D reconstruction.

4.1.3 The pinhole camera model

The pinhole camera model forms the basis for the majority of cameras and lenses that exhibit perspective projection. For the most part, depending on the required level of accuracy, only small adaptations are required. Central projection theory describes how an image is formed in a pinhole camera. In essence, a camera can be modelled as a central projection point C and an image plane. A ray can then be projected from a point in 3D space through the projection centre, resulting in an intersection with the imaging plane. This intersection can then be used to calculate the point at which an object at the world coordinates $[x\ y\ z]$ would appear on the image produced by the camera (Hartley and Zisserman 2004). The principle of central projection is shown in figure 4.1.

For a point $\mathbf{X}$ projected through a pinhole at $\mathbf{C}$, the coordinates must first be transformed into the camera coordinates $[x_c\ y_c\ z_c]$. This transformation can be broken down into two stages. First, a simple translation vector $\mathbf{T}$ is applied so that the object coordinates are relative to the camera centre of projection. Second, a rotation $\mathbf{R}$ is used to orientate the camera axis with the z-axis of the camera coordinate system. With the coordinates now in the correct frame, the intrinsic camera matrix $\mathbf{K}$ can be applied to determine the intersection point of the ray with the imaging plane. This projection is given by (Medioni and Kang 2004)

$$t\begin{bmatrix} u \\ v \\ 1 \end{bmatrix} = \mathbf{P}\begin{bmatrix} x \\ y \\ z \\ 1 \end{bmatrix} = \mathbf{K}\begin{bmatrix} \mathbf{R} & \mathbf{T} \\ 0 & 1 \end{bmatrix}\begin{bmatrix} x \\ y \\ z \\ 1 \end{bmatrix}, \tag{4.1}$$

where $[u\ v 1]$ is the intersection point with the imaging sensor, t is a scale factor and

$$\mathbf{P} = \mathbf{K}\begin{bmatrix} \mathbf{R} & \mathbf{T} \\ 0 & \mathbf{1} \end{bmatrix}. \tag{4.2}$$

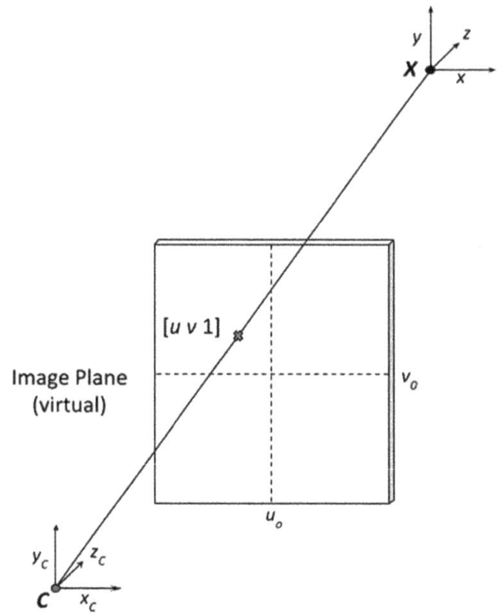

Figure 4.1. The basic principle of image plane intersection in a pinhole camera. The image plane is labelled as virtual because, although the actual imaging takes place behind the projection centre, the mathematics is simplified by using the virtual image plane.

The intrinsic camera matrix **K** depends on the internal properties of the camera being used. As a result, properties such as pixel size and the focal length of the lens have a significant impact on the final image. The general form of **K** is given by

$$\mathbf{K} = \begin{bmatrix} k_u & s & u_0 \\ 0 & k_v & v_0 \\ 0 & 0 & 1 \end{bmatrix}, \tag{4.3}$$

where k_u and k_v represent the focal lengths in terms of pixels in the u and v directions, respectively, u_0 and v_0 are the coordinates of the principle point and s is an additional term describing the skew of the image (Yang and Fang 2014). The principle point is the point at which the principle axis of the imaging system intersects with the imaging plane. Although it would typically be assumed that the principle point would lie at the centre of the image, small manufacturing errors in the camera body and lens can lead to a slight offset of the optical components. This offset results in the actual principle point being offset from the centre of the camera sensor. Although the offset may be small, it must still be accounted for.

For a pinhole camera, equation (4.1) perfectly describes the image. Unfortunately, with camera optical components, such as lenses, the simple linear relationship given by equation (4.1) no longer accurately describes the imaging process due to image distortions. Many of the cameras required to achieve sub-millimetre resolutions use optical components that result in such distortions. Despite the small alterations made to the pinhole model, it still forms the basis of a significant number of stereo photogrammetry methods.

4.1.4 Distortion modelling

Most recent advances in the pinhole model account for lens distortions within the model parameters. Using extensive models that include these distortions is critical when attempting to achieve accurate characterisation of the camera system. Therefore, it is important to fully understand the sources and effects of these distortions. In general, the distortions are parameterised through the additional terms δu and δv, as shown in figure 4.2.

There are three typical forms of distortion: radial, decentring and thin prism distortion. Radial distortion is a radially symmetric deformation about the principle point resulting in either a barrel or pin-cushion effect (Galantucci *et al* 2016). Decentring distortions are a result of the optical centre not being correctly aligned with the centre of the camera. This misalignment of the optical centre results in both radial and tangential distortions that must be included in the model. Thin prism distortion is a result of imperfections in the design and manufacturing of the camera lens and can be modelled as a thin prism being added to the optical system (Salvi *et al* 2002).

With all the potential sources of error, it is important to include enough parameters within the model such that it is flexible enough to fit the data. This concept has been used to calibrate long focal length macro lenses (Percoco *et al* 2015). With a long focal length lens, the distortions become significant and require consideration during both the calibration process and reconstruction. Through modelling up to sixth order radial and second order tangential distortion, Percoco and Sánchez Salmerón (2015) were able to digitise a tool part with a radial standard deviation of 8 μm over a 3.4° angle of view. Despite earlier models predicting a minimum angle of view of 10°, correct modelling allows lenses with smaller angles of view to be used, resulting in higher spatial resolutions (Stamatopoulos and Fraser 2011).

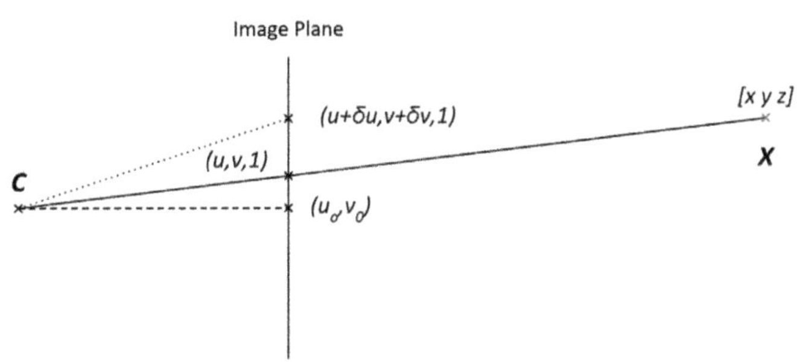

Figure 4.2. Lens distortion effect on the image plane intersection point. The observed intersection is now at $[u + \delta u, v + \delta v]$.

4.2 Characterisation and calibration

4.2.1 Camera characterisation

Characterisation is required in order to achieve accurate and reliable measurements. Camera characterisation aims to accurately determine the exterior orientation and intrinsic properties of the camera, effectively determining all the parameters described by the chosen camera model. Reconstructions are possible with un-calibrated cameras; however, no metric information can be obtained, making such applications unsuitable in a metrological context. Fortunately, the camera models discussed in the previous section provide a method of determining camera properties. By taking images of an object with known dimensions, it is possible to deduce the camera properties by fitting the image and object coordinates to a camera model. This method forms the basis of camera characterisation.

Camera characterisation methods fall into three basic categories: linear optimisation, non-linear optimisation and two-step techniques. Non-linear techniques are capable of dealing with models that exhibit non-linear relationships, such as lens distortions. The iterative nature of non-linear techniques mean that they can be slow and are not always guaranteed to produce a correct result (Luhmann *et al* 2016). Non-linear techniques rely on the minimisation of some function; as a result, an initial guess must be made that is reasonably close to the global minima. A significant deviation in initial parameters may result in a local minima being reached instead, producing an incorrect characterisation (Salvi *et al* 2002).

Linear techniques are much faster due to their more simplistic approach (Luhmann and Robson 2011). The linear method requires at least six corresponding points to work; however, by including many more points, a least-squares regression can be used to refine the parameters (Luhmann and Robson 2011). Although the linear method benefits from its simplicity and speed, it is incapable of taking into account lens distortions. There may also be some issues when extracting the camera properties from the transformation matrix as the linear method is not able to account for lens distortion (Salvi *et al* 2002).

Finally, the two-step technique is a combination of both the linear and non-linear techniques. A linear technique is first used to calculate some initial guess and fix some of the camera parameters (Lourakis and Deriche 1999). With a good initial guess, the non-linear technique can be used to improve the chances of reaching a global minimum. The two-step technique is perhaps the most widely used technique as it increases speed using linear methods, whilst still being able to account for image distortions (Salvi *et al* 2002).

4.2.2 Linear techniques

Linear techniques first require a definition for the transformation matrix that is being solved. Equation (4.1) can be re-written in the form

$$t\begin{bmatrix} u \\ v \\ 1 \end{bmatrix} = \begin{bmatrix} A_{11} & A_{12} & A_{13} & A_{14} \\ A_{21} & A_{22} & A_{23} & A_{24} \\ A_{31} & A_{32} & A_{33} & A_{34} \end{bmatrix} \begin{bmatrix} x \\ y \\ z \\ 1 \end{bmatrix}, \tag{4.4}$$

where **A** is the projection matrix (Luhmann and Robson 2011). Equation (4.4) can then be used to solve for u_i and v_i to give the collinearity equations

$$u = \frac{A_{11}x + A_{12}y + A_{13}z + A_{14}}{A_{31}x + A_{32}y + A_{33}z + A_{34}}, \tag{4.5}$$

$$v = \frac{A_{21}x + A_{22}y + A_{23}z + A_{24}}{A_{31}x + A_{32}y + A_{33}z + A_{34}}, \tag{4.6}$$

where the t term from equation (4.4) is cancelled, leaving only the camera properties and the 3D coordinates of the point being imaged (Luhmann and Robson 2011). The collinearity equations can then be used to define a linear relationship between the image and world coordinates and the vector matrix $\mathbf{A}'$ given by

$$\mathbf{QA}' = 0, \tag{4.7}$$

where

$$\mathbf{Q} = \begin{bmatrix} x_1 & 0 & \cdots & \cdots & x_n & 0 \\ y_1 & 0 & \cdots & \cdots & y_n & 0 \\ z_1 & 0 & \cdots & \cdots & z_n & 0 \\ 1 & 0 & \cdots & \cdots & 1 & 0 \\ 0 & x_1 & \cdots & \cdots & 0 & x_n \\ 0 & y_1 & \cdots & \cdots & 0 & y_n \\ 0 & z_1 & \cdots & \cdots & 0 & z_n \\ 0 & 1 & \cdots & \cdots & 0 & 1 \\ -u_1x_1 & -v_1x_1 & \cdots & \cdots & -u_nx_n & -v_nx_n \\ -u_1y_1 & -v_1y_1 & \cdots & \cdots & -u_ny_n & -v_ny_n \\ -u_1z_1 & -v_1z_1 & \cdots & \cdots & -u_nz_n & -v_nz_n \\ u_1 & v_1 & \cdots & \cdots & u_n & v_n \end{bmatrix}^T, \tag{4.8}$$

and

$$\mathbf{A} = \begin{bmatrix} A_{11} \\ A_{12} \\ A_{13} \\ A_{14} \\ A_{21} \\ A_{22} \\ A_{23} \\ A_{24} \\ A_{31} \\ A_{32} \\ A_{33} \\ A_{34} \end{bmatrix}, \tag{4.9}$$

and $\boldsymbol{n}$ is the number of points (Salvi *et al* 2002).

There are twelve unknown parameters in equation (4.7); as a result, a minimum of six corresponding points are required ($n \geqslant 6$). When n is greater than six, using the linear relation given in equation (4.7), it is possible to solve for $\mathbf{A}'$ using a least-squares regression method. The least-squares regression aims to minimise the function $\| \mathbf{QA}' \|^2$ with the constraint that $\| \mathbf{A}' \| = 1$. Through solving the minimisation of $\| \mathbf{QA}' \|^2$, the solution of $\mathbf{A}'$ is found to be the eigenvector of $\mathbf{Q}^\mathrm{T}\mathbf{Q}$ corresponding to the smallest eigenvalue (Medioni and Kang 2004).

Solving the eigenvector of $\mathbf{Q}^\mathrm{T}\mathbf{Q}$ to calculate $\mathbf{A}'$ provides a fast and simple method of camera characterisation (Lourakis and Deriche 1999).

As discussed earlier, linear characterisation methods are unable to account for distortions within images making them unsuitable for accurate characterisation. However, linear methods do offer a method for initial parameter estimation. The eigenvector method of parameter initialisation has been used to calculate initial extrinsic parameter and focal length estimations (Gu *et al* 2015). In order to make the linear method applicable, two assumptions are made: all coefficients of lens distortion are zero and the principle point is located at the exact centre of the image. Based on these assumptions, the extrinsic and focal length parameters can be extracted from $\mathbf{A}'$, and calculated using the eigenvector method described above (Zhang 1999). With the initial parameters calculated, the values can be further refined using a non-linear minimisation technique.

4.2.3 Non-linear methods

As linear techniques are incapable of modelling image distortions, other more complex, approaches must be considered. The methods of calibrating camera models that take into account the complex distortions that occur in many cameras are in general non-linear techniques.

For radial distortion, δu and δv are given by

$$\delta u = u(1 + k_1 r^2 + k_2 r^4 + k_3 r^6), \tag{4.10}$$

$$\delta v = v(1 + k_1 r^2 + k_2 r^4 + k_3 r^6), \tag{4.11}$$

where $r = \sqrt{(u^2 + v^2)}$ is the radial distance and k_1, k_2 and k_3 are the radial distortion coefficients. The effect of the various radial distortion coefficients is shown in figure 4.3, where k_1, k_2 and k_3 correspond to the second, fourth and sixth order contributions to the distortion.

Typically equations (4.10) and (4.11) are estimated up to the second or third term in order to have sufficient parameters, see Chen *et al* (2014) and Galantucci *et al* (2016) for examples. Equations (4.10) and (4.11) can then be combined with equations (4.5) and (4.6), respectively, to produce the distorted collinearity equations. With radial distortion now included, the camera model exhibits non-linear behaviour due to the addition of the radial distance term r.

Non-linear techniques are centred on the minimisation of some function $f(\tilde{\mathbf{x}}, \mathbf{m}_i, M_i)$, where $\mathbf{m}_i$ is the image coordinates of point i, $\mathbf{M}_i$ the physical coordinates and $\tilde{\mathbf{x}}$ is a vector containing all the chosen camera properties. Mathematically, this minimisation is expressed as

$$\text{minimise}_{\tilde{\mathbf{x}}} \sum_{i=1}^{N} \| f(\tilde{\mathbf{x}}, \mathbf{m}_i, \mathbf{M}_i) \|^2 \tag{4.12}$$

where $f(\tilde{\mathbf{x}}, \mathbf{m}_i, \mathbf{M}_i) = \mathbf{m}_i - \bar{\mathbf{m}}(\tilde{\mathbf{x}}, \mathbf{M}_i)$ and $\bar{\mathbf{m}}$ is the projection of the 3D coordinates $\mathbf{M}_i$ onto 2D image coordinates based on the camera properties $\tilde{\mathbf{x}}$ (Medioni and Kang 2004). The minimisation problem in equation (4.12) can be solved via established algorithms, such as the Levenberg–Marquardt (LM) algorithm. There are other algorithms capable of solving the minimisation problem shown in equation (4.12);

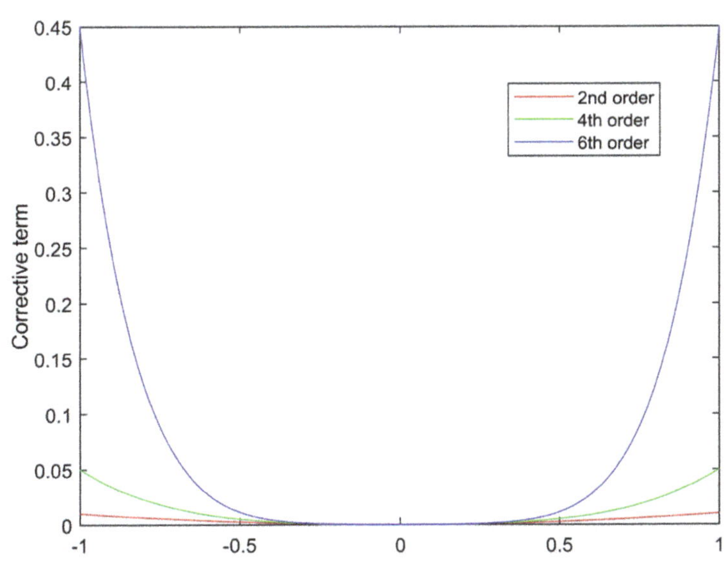

Figure 4.3. Representation of the effect of radial distortion of varying order. The legend refers to the maximum order of distortion that is taken into account.

however, the LM algorithm is the most suitable and widely used. The LM algorithm is an iterative process by which the variable vector is iteratively updated in the direction that minimises the cost function. For the case of camera characterisation, the variable to be changed is the camera properties vector $\tilde{\mathbf{x}}$. For iteration, n, the next value of $\tilde{\mathbf{x}}$ is given by $\tilde{\mathbf{x}}_{n+1} = \tilde{\mathbf{x}}_n + \Delta\tilde{\mathbf{x}}$, where $\Delta\tilde{\mathbf{x}}$ is calculated as

$$\Delta\tilde{\mathbf{x}} = (\mathbf{J}^{\mathrm{T}}\mathbf{J} + \lambda\mathbf{I})^{-1}\mathbf{J}^{\mathrm{T}}f(\tilde{\mathbf{x}}_n), \tag{4.13}$$

where $\mathbf{J}$ is the Jacobian and λ is a damping factor (Vaudrey and Klette 2009). The damping factor shifts the minimisation between a gradient decent and a Gauss–Newton method of minimisation. Reducing the damping factor makes the minimisation algorithm behave like the Gauss–Newton algorithm, suitable for when the reduction in the minimisation function is fast. However, the damping factor can be increased when the reduction in the residual slows down, resulting in an algorithm that behaves similarly to a gradient decent method. The variation in the damping factor ensures the LM algorithm approaches the global minima rather than becoming trapped in local minima.

Many current areas of research in close-range photogrammetry are focused on altering the LM algorithm to produce a more accurate characterisation. Gu *et al* (2015) provide an example of such an adaptation of the LM algorithm through the use of back projection. Equation (4.12) is essentially the sum of the residual vectors produced from the difference between the expected and observed image coordinates for given camera parameters. The method described by Gu *et al* instead calculates the residual as the difference between known 3D coordinates and the observed image coordinates projected back into 3D space. Conventional characterisation algorithms typically attempt to minimise the difference between the detected image point and the expected image point given by the projection model. However, by instead assuming that the 3D point lies on a plane, the intersection between the observed image point vector and the plane can be calculated. The difference between the known point on the plane and the observed image point vector intersection can then be minimised instead. Through the use of the back projection characterisation method, Gu *et al* were able to increase accuracies from 2 to 6 μm compared to the typical forward imaging process, achieving errors of 42 and 41 μm in the x- and y-directions, respectively (Gu *et al* 2015).

The nature of the LM algorithm allows it to be applied to any camera model, including the orthogonal projection model. Chen *et al* (2014) use the LM algorithm to calibrate a telecentric lens based camera model through the use of a 2D checkerboard artefact. The checkerboard artefact provides the known world points required for characterisation, and by measuring a known depth change of the artefact, the depth uncertainty is also determined. Through the use of the LM algorithm on the telecentric lens camera, a depth uncertainty of 2.9 μm was attained with a lateral resolution of 11 μm.

Algorithm efficiency also becomes an important factor when considering iterative methods. The Jacobean matrix is particularly time consuming to determine and reducing the computational cost would, therefore, be beneficial. Vaudrey and Klette (2009) outline a method that separates the intrinsic and extrinsic camera parameters and solves them. By splitting up the intrinsic and extrinsic parameters, the new form of the Jacobean matrix exhibits significant sparsity. The sparsity of the Jacobean matrix is a result of the majority of non-zero components lying on the diagonal. The sparsity of the new matrix allows it to be calculated more efficiently, particularly when calculating the inverse of $(\mathbf{J}\mathbf{J}^{\mathrm{T}} + \lambda\mathbf{I})$, as required in equation (4.13) (Vaudrey and Klette 2009).

4.2.4 Self-calibration

In some circumstances, it is difficult or impossible to know the geometry of a scene within a set of images. In the case of unknown scene geometry, the methods described in section 4.2.2 are no longer applicable. Rather than considering $\mathbf{M}_i$ as a set of fixed set of locations, self-calibration treats it as variable to be refined when minimising equation (4.12). Through the comparison of multiple images from cameras at different positions, it is possible to simultaneously alter both the camera parameters $\tilde{\mathbf{x}}$ and corresponding point coordinates $\mathbf{M}_i$. The self-calibration minimisation problem requires a slight adaptation to equation (4.13) to give

$$\text{minimise}_{\tilde{\mathbf{x}}} \sum_{i=1}^{N} \sum_{j=1}^{M} \left\| \mathbf{m}_{i,j} - \bar{\mathbf{m}}\left(\tilde{\mathbf{x}}_i, \mathbf{M}_j\right) \right\|^2, \tag{4.14}$$

where $\mathbf{m}_{i,j}$ is the image coordinate of point i produced by camera j (Zhang 1999). The minimisation problem described in equation (4.14) is solved in exactly the same way as described in section 4.2.2 and effectively calibrates the camera and reconstructs the 3D scene at the same time.

4.3 System calibration

(Reproduced from (Sims-Waterhouse *et al* 2020).)

The strength of photogrammetry lies in its capacity to achieve reconstruction without relying on any additional information other than the images of the object being measured. Despite the advantages of reconstructing geometries from images, this can only be achieved at the loss of scale information (Fragione *et al* 2019). Additionally, the complex and non-linear process of characterising the system properties and triangulation of the sample form makes the propagation of error through the system difficult to calculate (Lavecchia *et al* 2018). The lack of standardised methodologies for scaling and consequent lack of traceability, plus a poor understanding of the uncertainty of photogrammetry measurements, are significant hurdles in the widespread use of photogrammetry as a coordinate measurement system.

4.3.1 Arbitrary scale

The solution to the arbitrary scale of reconstructions is a key area of research into the metrological applications of photogrammetry methods. Typically, some physical property of the measurement system and/or the sample must be fixed and calibrated in order for measurements to be appropriately scaled (Luhmann 2010). For example, a calibrated artefact can be placed within the measurement volume of the photogrammetry system allowing the correct scale factor to be applied after the measurement is taken. Although this approach is easily applied, it greatly limits the working volume of the system and requires a calibrated artefact to be available for all measurements.

Another common approach, in the case of a multi-camera system, is to fix the physical distance between the multiple cameras within the system and apply that physical property to the measurement reconstruction (Sun *et al* 2018). This is a typical approach in many optical coordinate measurement techniques, particularly in fringe projection measurement systems (Duan *et al* 2018). The calibrated artefact must be measurable by the system, as well as traceably calibrated to a sufficient level of accuracy for the system under consideration. For contact CMS, gauge blocks are easily used and widely available, whilst being traceably calibrated to high accuracies (<0.06 μm for grade K gauge blocks, ISO 3650 (ISO 1998)). However, active optical measurement systems are typically not able to measure steel gauge blocks due to their highly specular surfaces (Li *et al* 2017). Instead, 2D patterned artefacts, such as checkerboards or dot patterns, are typically used (Zhang 1999). While checkerboard artefacts may be effective for the calibration of optical CMS, they are difficult to calibrate to high accuracies (<1 μm) and can be difficult to use for non-experts. Ball bars are a widely used approach for the calibration of optical CMSs (Sims-Waterhouse *et al* 2017), with the sphere-to-sphere distances providing the scaled length. Similar to 2D patterned artefacts, ball bars are difficult to use for calibration as they exhibit little observable surface texture for photogrammetry measurements, although texture projection methods have been shown to make this feasible (Sims-Waterhouse *et al* 2017).

An alternative method to effectively calibrate optical CMSs is to use a combination of self-calibration and direct measurement of a single calibrated length (Sims-Waterhouse *et al* 2017). Self-calibration does not require a calibrated object to effectively characterise the optical system up to some arbitrary scale factor. By then measuring a calibrated length, such as a gauge block, the arbitrary scale factor of the system can be fixed. Although gauge blocks are typically difficult to measure with optical CMS, this issue is simplified as only a single measurement is required—in fact, limiting the required number of measurements ensures that the gauge block can be measured in the most optimal position, reducing specular reflection and optimising the measurement accuracy. This methodology is more easily implemented as an intermittent calibration procedure for optical CMS, whilst only requiring widely available calibration artefacts.

4.4 Image acquisition

4.4.1 Depth of field

In theory, there will only be a single plane that will be in focus when taking an image. In practice there will, in fact, be a range of points that appear to be in focus. The range over which points appear to be in focus is the depth of field (Percoco and Sánchez Salmerón 2015). The ability of a system to detect whether or not a point is in focus is determined by the circle of confusion, defined as the diameter of the largest blurred spot that appears as a point to the human eye. For a 35 mm aperture, the circle of confusion is conventionally set to 0.03 mm; differences in viewing conditions and individual differences can, however, lead to small changes to this value (Percoco and Sánchez Salmerón 2015). The principle of the depth of field is shown in figure 4.4.

As figure 4.4 suggests, the circle of confusion fixes the minimum and maximum distances for which a point will be in focus for a particular f-number. By changing the f-number, and hence the aperture size, the ray lines in figure 4.4 will either be squeezed inwards or spread out. As the circle of confusion is a constant value, reducing the angle of the light rays will increase the depth of field and decrease it for increased angle. This effect is highlighted by the equation for depth of field given by

$$d = 2NC\left(1 + \frac{1}{M}\right), \tag{4.15}$$

where N is the numerical aperture, C is the diameter of the circle of least confusion and M is the magnification.

In order to ensure diffraction effects do not limit the imaging system, a lower f-number should be chosen. Choosing a low f-number directly leads to a decreased depth of field, meaning a reduced working range. The trade-off between resolution and depth of field is purely a result of optical effects and, therefore, is difficult to overcome. For many applications, simply choosing the optimal f-number for the sample dimensions would be adequate. However, in order for photogrammetry to be useful from micro-scale features, this trade-off between resolution and depth of field must be overcome.

Gallo *et al* (2014) provide a method of overcoming the small depths of field produced by macro lenses through the use of image stacking. By taking many images at varying focus distances, a stack of images can be produced for which each pixel is in focus in at least one image. A fusion algorithm can then be applied to the

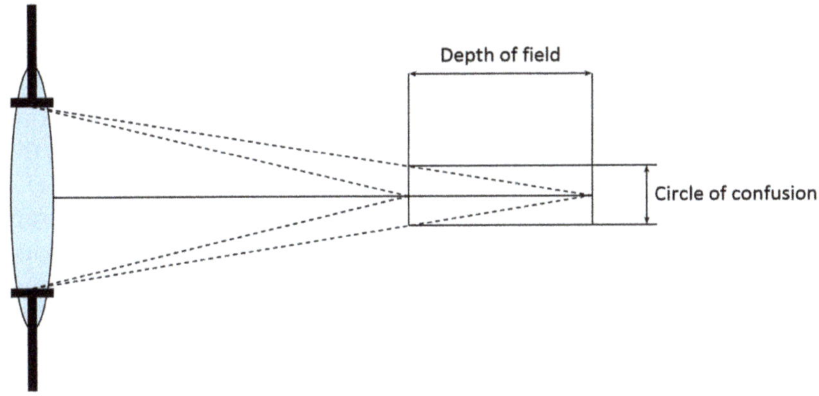

Figure 4.4. Principle of the depth of field for a lens and aperture. The dashed lines correspond to the light rays from the closest and furthest points in focus.

stack of images in order to produce a singular image for which all points are in focus. Gallo *et al* were able to measure step heights of 20–100 μm to an uncertainty of 10 μm with the multi-focus image stacking method at a magnification of one. Compared to Percoco and Sánchez Salmerón (2015), Gallo *et al* were able to achieve similar accuracies at half the magnification and with a significantly larger sampling area. The downside to the image stacking method is the significant image acquisition times; complex objects took around 2 h to image due to the methodology used.

Another method of overcoming depth of field constraints is to alter the optics and remove the issue entirely. Plenoptic cameras (light field cameras) use a different method of image acquisition for which depth of field and focus are no longer limiting factors (Zeller *et al* 2014). By placing an array of micro-lenses between the imaging sensor and optical components the entirety of the light field can be detected, not just intensity and position. Plenoptic cameras are able to detect the position, incoming angle, intensity and colour of incoming light rays allowing for image reconstruction after acquisition. The current use of plenoptic cameras for depth mapping is limited to single camera methods using only the light field data (Weimer *et al* 2014, Zeller *et al* 2014, Rangappa *et al* 2019). The level of accuracy achieved by plenoptic cameras is still relatively low and their current resolution is significantly lower than regular DSLR cameras.

Methods of fusing the data produced by plenoptic cameras with photogrammetry methods have also been demonstrated (Sims-Waterhouse *et al* 2017). In this work, a consumer plenoptic camera was used to capture images with an extended depth of field and the part geometry was reconstructed using photogrammetry methods. As expected, the low-resolution images reflected on the poor quality of the photogrammetry data. However, the plenotpic camera also allowed for the metric size of the part to be extracted without the use of an additional calibration step.

4.4.2 Imaging procedure optimisation

There are two main concerns when view planning for an optical CMS: where should the camera views be placed relative to the object, and how many camera views are required to satisfy object coverage and reconstruction accuracy requirements in a reasonable amount of time (Olague 2001). As can be seen in figure 4.5, complex parts exhibit a range of features that can be difficult to measure accurately. For example, high aspect ratio features and partially occluded regions can be difficult to measure due to measurements requiring multiple lines of sight. Carrivick *et al* (2016) found that tens to hundreds of images are typically required for accurate reconstruction, depending on object complexity, desired reconstruction accuracy and reconstruction time constraints. In a multi-camera network, as the number of cameras increases, so does the computational cost and bandwidth required; therefore, many state-of-the-art systems attempt to minimise the number of cameras in the network (Xing *et al* 2017).

Early work in this field explored the development of the ‘configuration of sensors’ (CONSENS) system (Mason and Grün 1995); this used expert knowledge to

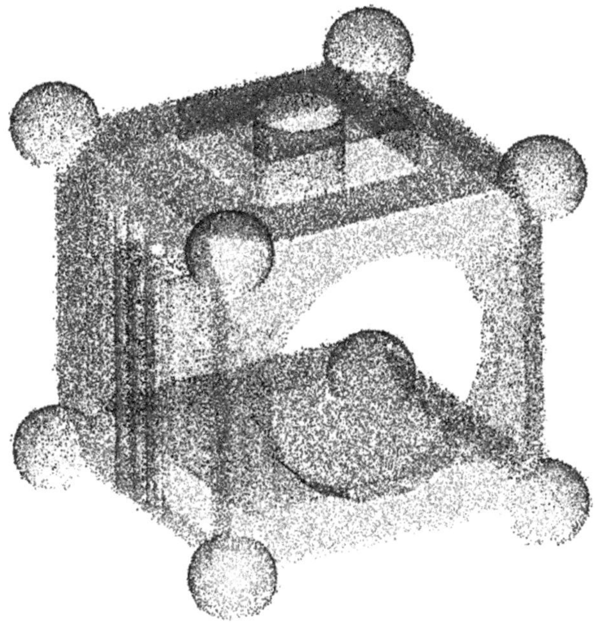

Figure 4.5. Point cloud of a complex part produced by a photogrammetry system.

conduct view planning from the CAD data of an object. The 'evolving position of cameras' (EPOCA) system was developed for designing imaging networks for robotic systems. EPOCA did not rely on expert knowledge and instead used genetic algorithms to optimise the imaging network. More recently, the imaging network designer (IND) system has been developed (Ahmadabadian *et al* 2014); this uses a four-step algorithm of datum definition, optimum distance calculation, viewpoint generation, and viewpoint clustering and selection. IND relies on initial object geometry, which can be obtained using a structure from a motion or fringe projection technique. Rangel *et al* (2019) uses evolutionary algorithms to optimise coverage and viewpoint redundancy in a multi-sensor measurement network. Some approaches attempt to iteratively improve on a sparse initial network by adding more camera views *in situ*; however, this approach requires a reasonable initial network and *in situ* computation times that could hamper measurement speed. For example, Erat *et al* (2019) present a real-time online variant of a view planning approach that accepts incoming views to the network if this view improves coverage of the scene, based on a predefined coverage metric.

In a recent work by Eastwood *et al* (2020), a combination of CNNs and camera pose optimisation was proposed to optimise the photogrammetry measurement procedure. Eastwood *et al* showed that by minimising an objective function that characterises the camera coverage of a particular geometry, an optimal set of camera locations can be generated. Substantial reduction in the measurement precision could be achieved with a low number of images when using optimised locations compared to uniform distributions. In addition to the camera location optimisation, Eastwood *et al* also proposed the use of CNNs to determine the pose of a sample to be measured prior to a measurement cycle. The use of CNNs and machine learning in measurement technologies is discussed in more detail in chapter 6. The outcome of the work suggested by Eastwood *et al* is a system capable of performing an automated measurement with optimal measurement accuracy without the requirement of a skilled user.

4.5 Summary

In summary, photogrammetry has seen improvement in speed and accuracy over the last decades. However, the application of photogrammetry in form and coordinate measurement is still immature, with research still required to fully utilise the technique. However, the potential benefits of photogrammetry are substantial. The combination of photogrammetry with smart algorithms and machine learning to produce fully automated systems is of particular importance.

References

Bergeles C, Pratt P and Merrifield R 2014 Multi-view stereo and advanced navigation for transanal endoscopic microsurgery *Int. Conf. Medical Image Computing and Computer-Assisted Intervention* pp 332–29

Carrivick J L, Smith M W and Quincey D J 2016 *Structure from Motion in the Geosciences* (New York: Wiley)

Chen Z, Liao H and Zhang X 2014 Telecentric stereo micro-vision system:Calibration method and experiments *Opt. Lasers Eng.* **57** 82–92

Duan X, Wang C, Wang J and Zhao H 2018 A new calibration method and optimization of structure parameters under the non-ideal condition for 3D measurement system based on fiber-optic interference fringe projection *Optik* **172** 424–30

Eastwood J, Zhang H, Isa M, Sims-Waterhouse D, Leach R K and Piano S 2020 Smart photogrammetry for three-dimensional shape measurement *Proc. SPIE* **11352** 11252A

Erat O, Hoell M, Haubenwallner K, Pirchheim C and Schmalstieg D 2019 Real-time view planning for unstructured lumigraph modeling *IEEE Trans. Vis. Comput. Graph.* **25** 3063–72

Fragione A, Sanchez Salmeron A J, Modica F and Percoco G 2019 Multi-step approach for automated scaling of photogrammetric micro-measurements *Int. J. Adv. Manuf. Tech.* **102** 747–57

Fraser C, Cronk S and Hanley H 2008 Close-range photogrammetry in traffic incident management *ISPRS Cong.* 3–6

Galantucci L M, Pesce M and Lavecchia F 2016 A powerful scanning methodology for 3D measurements of small parts with complex surfaces and sub millimeter-sized features, based on close range photogrammetry *Precis. Eng.* **43** 211–9

Gallo A, Muzzupappa M and Bruno F 2014 3D reconstruction of small sized objects from a sequence of multi-focused images *J. Cult. Herit.* **15** 173–82

Gu F, Zhao H, Ma Y and Bu P 2015 Camera calibration based on the back projection process *Meas. Sci. Technol.* **26** 125004

Hartley R and Zisserman A 2004 *Multiple View Geometry in Computer Vision* 2nd edn (Cambridge: Cambridge University Press)

ISO 3650 1998 *Geometrical Product Specifications (GPS)—Length Standards—Gauge Blocks* (Geneva: International Organization for Standardization)

Lavecchia F, Guerra M G and Galantucci L M 2018 Performance verification of a photogrammetric scanning system for micro-parts using a three-dimensional artifact: adjustment and calibration *Int. J. Adv. Manuf. Technol.* **96** 4267–79

Li J, Ren H, Luo P, Gao X and Wang Z 2017 Specular reflection compensation in homography fringe projection profilometry *Optik* **140** 413–22

Lourakis M and Deriche R 1999 Camera self-calibration using the singular value decomposition of the fundamental matrix: from point correspondences to 3D measurements *Institut National de Recherche en Informatique et en Automatique* No 3748

Luhmann T 2010 Close range photogrammetry for industrial applications *ISPRS J. Photogramm.* **65** 558–69

Luhmann T, Fraser C and Maas H G 2016 Sensor modelling and camera calibration for close-range photogrammetry *ISPRS J. Photogramm* **115** 37–46

Luhmann T and Robson S 2011 *Close Range Photogrammetry: Principles, Techniques and Applications* (Dunbeath, UK: Whittles)

Mason S O and Grün A 1995 Automatic sensor placement for accurate dimensional inspection *Comput. Vis. Image. Underst.* **61** 454–67

Medioni G and Kang S B 2004 *Emerging Topics in Computer Vision* (Englewood Cliffs, NJ: Prentice Hall)

Miura Y, Nakanishi S, Higuchi E, Takamasu K, Abe M and Sato O 2019 Comparative evaluation of estimation of hole plate measurement uncertainty via Monte Carlo simulation *Precis. Eng.* **56** 496–505

Möller C, Schmidt H C, Shah N H and Wollnack J 2016 Enhanced absolute accuracy of an industrial milling robot using stereo camera system *Proc. Technol.* **26** 389–98

Olague G 2001 Autonomous photogrammetric network design using genetic algorithms *Applications of Evolutionary Computing* (Lecture Notes in Computer Science vol 2037) (Berlin: Springer) pp

Percoco G, Lavecchia F, Sánchez and Salmerón A J 2015 Preliminary study on the 3D digitization of millimeter scale products by means of photogrammetry *Proc. CIRP* 33 257–62

Percoco G, Sánchez and Salmerón A J 2015 Photogrammetric measurement of 3D freeform millimetre-sized objects with micro features: an experimental validation of the close-range camera calibration model for narrow angles of view *Meas. Sci. Tech.* **26** 95203

Rangappa S, Matharu R and Petzing J 2019 Establishing the performance of low-cost Lytro cameras for 3D coordinate geometry measurements *Mach. Vision Appl.* **30** 615–27

Rangel E O, Costa D G and Loula A 2019 On redundant coverage maximization in wireless visual sensor networks: evolutionary algorithms for multi-objective optimization *Appl. Soft Comput.* **82** 105578

Remondino F, Baltsavias E, El-Hakim S, Picard M and Grammatikopoulos L 2008 Image-based 3D modeling of the Erechteion, Acropolis of Athens *ISPRS Ann.* **37** 1083–92

Salvi J, Armanguã X and Batlle J 2002 A comparative review of camera calibrating methods with accuracy evaluation *Pattern Recognit.* **35** 1617–35

Sims-Waterhouse D, Isa M, Piano S and Leach R K 2020 Uncertainty model for a traceable stereo-photogrammetry system *Precis. Eng.* **63** 1–9

Sims-Waterhouse D, Piano S and Leach R K 2017a Verification of micro-scale photogrammetry for smooth three-dimensional object measurement *Meas. Sci. Technol.* **28** 055010

Sims-Waterhouse D, Piano S and Leach R K 2017b Fusion of light-field and photogrammetric surface form data *Proc. SPIE* 10373 1037304

Stamatopoulos C and Fraser C 2011 Calibration of long focal length cameras in close range photogrammetry *Photogramm. Rec.* **26** 339–60

Sun P, Lu N G, Dong M L, Yan B X and Wang J 2018 Simultaneous all-parameters calibration and assessment of a stereo camera pair using a scale bar *Sensors* **18** 1–19

Vaudrey T and Klette R 2009 Computer analysis of images and patterns *Proc. IEEE Int. Conf. Comput. Vis.* 5702 541–8

Wang X, Liu J, Liu S, Jin P, Wu T and Wang Z 2018 Accurate radius measurement of multi-bend tubes based on stereo vision *Measurement* **117** 326–38

Weimer D, Thamer H, Fellmann C, Lütjen M, Thoben K D and Scholz-Reiter B 2014 Towards 100% *in situ* 2D/3D quality inspection of metallic micro components using plenoptic cameras *Proc. CIRP* 17 847–52

Xing Z, Yu J and Ma Y 2017 A new calibration technique for multi-camera systems of limited overlapping field-of-views *IROS* pp 5892–9

Yang X and Fang S 2014 Effect of field of view on the accuracy of camera calibration *Optik* **125** 844–9

Zeller N, Quint F and Stilla U 2014 Calibration and accuracy analysis of a focused plenoptic camera *ISPRS II-3* pp 205–12

Zhang Z 1999 Flexible camera calibration by viewing a plane from unknown orientations *Proc. ICCV 1* pp 666–73

IOP Publishing

Advances in Optical Form and Coordinate Metrology

Richard Leach

Chapter 5

Digital fringe projection profilometry

Rui Chen, Jing Xu and Song Zhang

5.1 Introduction

This chapter introduces the state-of-the-art technologies in fringe projection profilometry (FPP), with a primary focus on those advancements made in the digital FPP field within the past decade. We divide the chapter into sections based on technological stages, from fringe generation, fringe analysis and phase unwrapping, to calibration. We also discuss recent advancements on high-speed FPP techniques and the effort made towards automation. The introduction of machine learning techniques will be discussed within each subsection, if appropriate. We conclude this chapter by presenting two application cases where fully integrated solutions offer opportunities for optimisation that would not be possible without collaborating with domain experts from other disciplines.

5.2 Fringe pattern generation methods

FPP is a method that performs three-dimensional (3D) shape measurement by projecting sinusoidally varying structured patterns (i.e. fringe patterns) and analysing the captured patterns scattered by the scene for 3D surface reconstruction. Recently, the fringe pattern definition has been generalised by removing the sinusoidal requirement because new analysis methods can analyse non-sinusoidal pattern(s) in a similar manner for 3D surface reconstruction (Xu and Zhang 2020). As such, more fringe generation methods can be used for FPP. This section covers conventional fringe generation techniques as well as some recently developed ones.

5.2.1 Conventional fringe generation methods

Fringe patterns were historically generated through one of the following means (Xu and Zhang 2020):

doi:10.1088/978-0-7503-2524-0ch5

- *Interference.* A sinusoidal fringe pattern is generated by superposing two laser beams with the same wavelength. This method has been extensively employed in laser interferometry systems (Harihara 2006).
- *Physical grating projection.* A sinusoidal fringe pattern is generated by projecting patterns created on a grating (Anderson and Porter 1929) or printed on a transparency film (Wust and Capson 1991).
- *Moiré.* A fringe pattern is generated by the Moiré effect: an opaque ruled pattern with transparent gaps is overlaid on another similar pattern (Rowe and Welford 1967).
- *Digital video projection.* A fringe pattern is generated using a video projector to project a computer generated fringe pattern (Zhang 2010). This method is often called the *digital fringe projection or DFP* technique. The flexibility and speed of DFP techniques have enabled rapid developments in the field. As such, this chapter primarily focuses on DFP techniques.

DFP techniques conventionally use a digital video projector to project computer generated 8 bit sinusoidal fringe patterns for 3D shape measurement. However, all commercial video projectors are non-linear (for example, the intensity response from an input image to an output image is non-linear) to accommodate the non-linear human perception system. Therefore, the computer generated sinusoidal patterns must be appropriately distorted to create ideal sinusoidal patterns for high-quality 3D shape measurement (Zhang 2015). Compared to liquid crystal displays (LCDs), digital light processing (DLP) projection techniques have the advantages of higher image contrast, higher pixel accuracy and better pixel uniformity. Unlike LCD projection systems that project red, green and blue (RGB) channels simultaneously, DLP projection systems project RGB channels sequentially. The unique projection mechanism of DLP technology has enabled the development of high-resolution real-time 3D shape measurement (Zhang and Huang 2006, Zhang and Yau 2006). However, it is not simple to reproduce such a high-speed and high-accuracy 3D shape measurement system because DLP employs the time modulation principle for 8 bit grayscale image generation (Hornbeck 1997) that requires accurate synchronisation between the projector and the camera. Furthermore, the refresh rate of DLP projectors is limited to approximately 120 Hz, a speed bottleneck that is difficult to break without further innovations (Zhang 2018).

5.2.2 Digital binary defocusing techniques

In parallel, binary coding methods (Salvi *et al* 2010) have been developed in the computer vision community, with some being able to achieve real-time performance (Rusinkiewicz *et al* 2002). DLP technology naturally produces binary images at a much higher frame rate than grayscale images (for example, tens of kilohertz against hundreds of hertz), and thus the binary coding methods are a better fit for such a platform to achieve high measurement speeds. However, the achievable spatial resolution of the binary coding methods is much lower than that for fringe projection methods. Digital binary defocusing technology (Lei and Zhang 2009) achieved the same spatial resolution as the fringe projection method and simultaneously took advantage of the binary projection mechanism of DLP systems. The DLP development kits enabled speed breakthroughs, achieving kilohertz rates of 3D shape measurement (Gong and Zhang 2010, Li and Zhang 2018, Zhang *et al* 2010, Zuo *et al* 2018). However, it is difficult for the simple binary defocusing method to achieve a similar quality of measurement, in particular when a small number of phase-shifted patterns is used (Su *et al* 1992).

To overcome the limitations of the single binary coding technique and improve measurement quality, researchers have developed various optimisation methods that include 1D (Ayubi *et al* 2010, Wang and Zhang 2010, Zuo *et al* 2012), 2D area (Dai *et al* 2014, Li and Zhang 2017, Lohry and Zhang 2013, Lu *et al* 2019, Sun *et al* 2015, Wang and Zhang 2012, Xiao and Li 2017, Xu *et al* 2017) and 3D (2D area + time) (Ajubi *et al* 2011, Silva *et al* 2017, Wang *et al* 2017, Zhu *et al* 2016) approaches. In addition, an optimisation method using multiple levels of grayscale values was developed (Zuo *et al* 2013).

Working on DLP platforms, binary defocusing techniques have enabled numerous breakthroughs in the field: the combined flexibility, accuracy, speed and ease of use are unparalleled. However, as the core of DLP technology uses a digital micromirror device (DMD) that is made of silicon, this limits the spectral range of light that can be used, as well as the power of the light being projected. Therefore, new projection technologies are needed to overcome these limitations to address applications that are not possible with DLP platforms.

5.2.3 New fringe pattern generation methods

To overcome the limitations of DLP technology, projectors using light emitting diode (LED) arrays were developed to create fringe patterns for high-speed 3D shape measurement (Fujigaki *et al* 2013, Heist *et al* 2014). However, LED array projectors still have limited power that can be generated for a reasonable package size. Heist *et al* (2016) further developed a projector with a mechanically rotating wheel. The rotating wheel has open and closed slots representing 1 s and 0 s of binary structured patterns. By appropriately defocusing the light from the projection lens system, a sequence of periodical sinusoidal fringe patterns are generated for 3D reconstruction. Since the mechanical projector does not use silicon-based devices for pattern generation, the light spectrum can be substantially extended to be used for applications such as 3D thermal imaging (Brahm *et al* 2016). Also, since the rotating wheel can be manufactured with metal or other materials that can tolerate more light power, such a technique successfully addresses the limitation of the DLP projection technique. However, the achieved 3D shape measurement speed is not as high as that using a DLP platform. To achieve higher 3D shape measurement speeds, Hyun *et al* (2018) developed a system that achieved 10kHz 3D shape measurement regardless of the number of phase-shifted fringe patterns required for each 3D reconstruction. Figure 5.1 shows a fast 3D shape measurement system with a mechanical projector and the corresponding measurement results. However, this system requires accurate synchronization between the mechanical projector and the

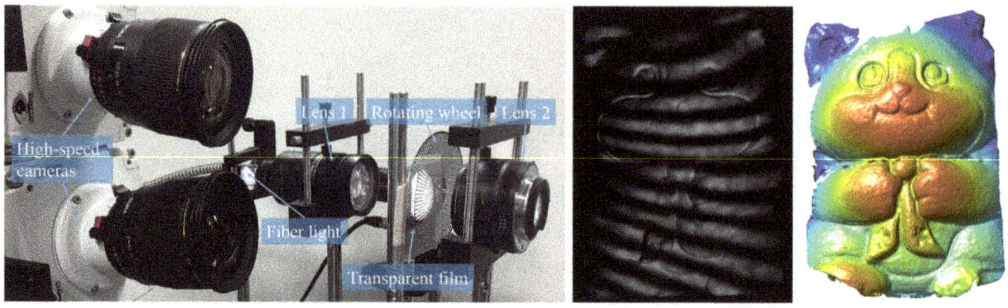

(a) (b) (c)

Figure 5.1. Fringe projection system using a mechanical projector: (a) photograph of the 3D shape measurement system with the mechanical projector; (b) phase-shifted fringe patterns captured by one of the cameras; and (c) 3D shape of the object shown in (b). Adapted with permission from Hyun *et al* (2018). Copyright 2918 The Optical Society.

camera which is not easily realised. Furthermore, the flexibility and accuracy of a mechanical projector is still not comparable to DLP projectors; as such, new developments are needed to further advance this field.

5.3 Fringe analysis

The second step in FPP is to retrieve the carrier phase from the fringe pattern(s) through *fringe analysis*. Fringe analysis essentially processes the fringe pattern(s) to obtain the phase $\phi(x, y)$ information. Numerous fringe analysis techniques, including Fourier transform (FT) and phase-shifting methods, have been developed and employed for many practical applications. The availability of machine learning tools has recently enabled new fringe analysis methods that could address some long-standing challenges in the fringe analysis field. This section overviews these techniques.

5.3.1 Conventional fringe analysis techniques

Fourier transform profilometry (FTP) methods analyse one fringe pattern to retrieve the phase using the Fourier transform (Takeda and Mutoh 1983). FTP has been successfully employed in numerous applications (Takeda 2013). However, FTP can recover accurate phase information only when the surface geometry is continuous and smooth and the surface texture varies moderately (Qian 2010). These constraints limit its practical applications. Therefore, various methods have been developed to improve the performance of single-pattern FTP methods, including using the windowed Fourier transform (Kemao 2004, Kemao *et al* 2008, Kemao 2015), multi-scale windowed Fourier transform (Zhong and Zeng 2007), DC signal elimination (Chen *et al* 2005), Hilbert transform (Gdeisat *et al* 2016, Hyun and Zhang 2020, Luo *et al* 2016) and wavelet transform (Abid *et al* 2007, Gdeisat *et al* 2006, Gdeisat *et al* 2009, Zhong and Weng 2004). Although these advanced methods can improve the phase retrieval accuracy to various degrees, they still have some fundamental problems if noise is large. Therefore, researchers have also proposed

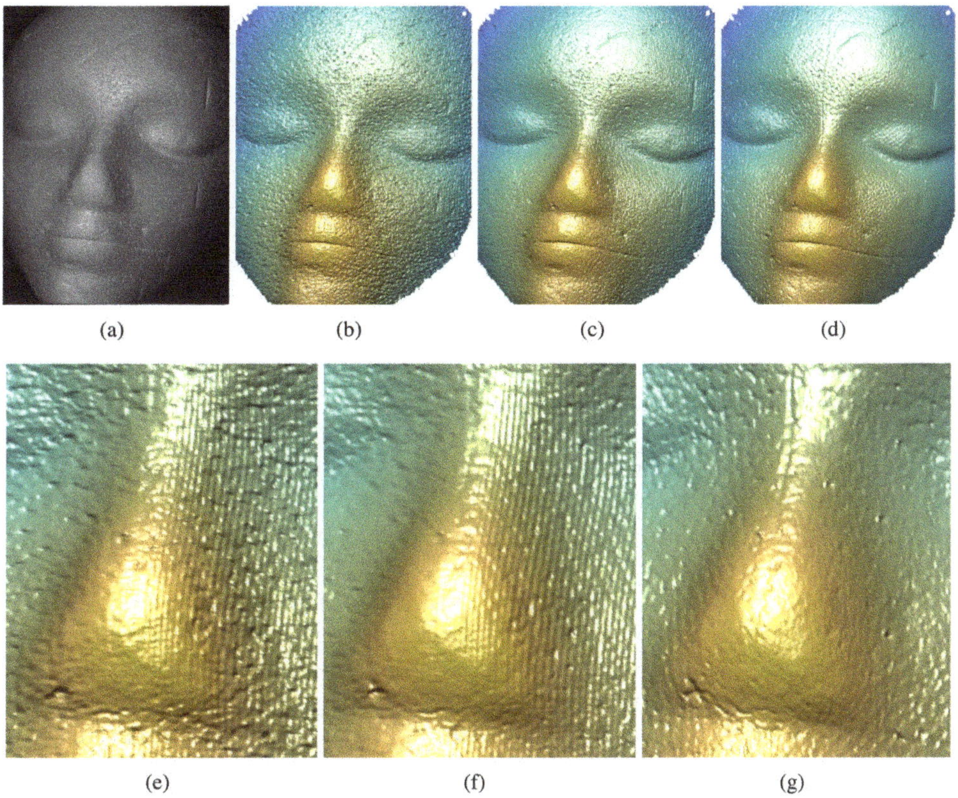

Figure 5.2. Measurement results using different fringe analysis techniques (Zhang 2018): (a) photograph of the object; (b) 3D result using the FTP method (Takeda and Mutoh 1983); (c) 3D result using the modified FTP method (Guo and Huang 2008); (d) 3D result using the three-step phase-shifting method; (e)–(g) close-up views of the results shown in (b)–(d). Reprinted from Zhang (2018), copyright 2018, with permission from Elsevier.

methods to improve the measurement quality by projecting one additional pattern to reduce the impact of the DC signal and enhance the carrier signal (Guo and Huang 2008, Li *et al* 1990, Zhang and Yau 2007). These methods have improved performance, albeit at the cost of capturing and processing another image. FTP, after all, requires at least a small region of the image to recover phase for 3D shape measurement, limiting its capability for complex surface measurements.

Phase-shifting profilometry (PSP) methods retrieve the phase pixel by pixel from a sequence of fringe patterns with known phase shifts. Because the phase of each pixel is recovered independently, rather than using neighborhood information as with FTP, PSP methods usually have better accuracy and robustness to surface geometry and changes in environmental illumination, and thus have been extensively employed in optical metrology. The underlying assumption of any PSP method is that the target object stays at least quasi-static during the projection of multiple fringe patterns, such that each pixel corresponds to the same surface point across all the captured images. The assumption could be violated in dynamic measurement applications, when motion artefacts are introduced. Even though some algorithms

(Liu *et al* 2018, Wang *et al* 2018) have been developed to reduce motion introduced artefacts, they cannot completely eliminate motion introduced error without improved hardware.

Figure 5.2 shows an example of a measurement of the same object with different fringe analysis techniques. This comparison of results shows that with the use of an increased number of fringe patterns, the measurement quality tends to improve, even for the static and geometrically simple object shown in this example.

5.3.2 Machine learning enhanced fringe analysis methods

In the past decades, machine learning has shown great success and significant improvement compared with traditional techniques for various tasks, including image classification (Krizhevsky *et al* 2012, Simonyan and Zisserman 2014), object detection (Girshick 2015, Ren *et al* 2015) and multi-view stereo (Chen *et al* 2019, Yao *et al* 2018) (see chapter 6). By using learned features from a large volume of data, machine learning techniques can extract useful information from input data and are robust to data defects. Therefore, researchers have begun to apply machine learning to fringe analysis to improve phase retrieval accuracy and resistance.

Feng *et al* (2019) proposed the use of convolutional neural networks (CNNs), a type of deep learning technique widely used in image processing, for single fringe pattern analysis. CNNs perform a convolution operation in the local neighborhood of an image and extract context information. By using multiple layers, CNNs can derive from low-level features (for example, edges, corners) to high-level features (for example, object category, object overall structures). Feng *et al* (2019) first used N-step phase-shifting to create accurate ground truth phase maps for network training, then trained two separate networks to predict intermediate results: the background image, the numerator and the denominator for phase calculation. Experiments demonstrate that learning-based fringe analysis can retrieve the phase more accurately than the traditional FTP and windowed FTP methods. One limitation of machine learning methods is that a large amount of data is required for network training to achieve accurate phase retrieval. These data are difficult and time-consuming to acquire in numerous practical applications where scenes may be dynamically changing. Shi *et al* (2019) addressed this problem by training the network with enhanced data and denoised data simultaneously and cropping the original input fringe patterns into overlapped small patches. Further, Van der Jeught and Dirckx (2019) proposed the use of CNNs to directly predict the height map from a single fringe pattern, which replaces the phase retrieval step entirely, but they only verified their method on simulated data.

Machine learning has shown promise in fringe analysis, yet its potentials and ultimate limitations have yet to be fully explored. More research is necessary to better understand the power and the applicability of machine learning tools.

5.4 Phase unwrapping

The phase $\phi(x, y)$ value obtained from a fringe analysis technique typically ranges from $-\pi$ to $+\pi$ with 2π discontinuities, which is often called the *wrapped phase*. 2π

discontinuity problems need to be addressed before 3D reconstruction. The process of removing 2π discontinuities in the wrapped phase map is called *phase unwrapping*. Phase unwrapping essentially determines the *fringe order*, an integer number $k(x, y)$ of 2πs to be added to the phase $\phi(x, y)$. Conventional phase unwrapping algorithms can be classified into two major categories: spatial phase unwrapping and temporal phase unwrapping. Recently, researchers in the FPP community have developed other novel phase unwrapping algorithms to overcome some drawbacks of the conventional phase unwrapping algorithms. Once again, machine learning tools provide new perspectives for phase unwrapping. A thorough survey on absolute phase unwrapping is given in Zhang (2018) and will be not be covered in detail in this section. This section will include some conventional phase unwrapping methods and some newly developed ones.

5.4.1 Conventional phase unwrapping methods

As discussed earlier, conventional phase unwrapping methods typically refer to spatial and temporal phase unwrapping. *Spatial phase unwrapping* methods determine the fringe order by referring the phase value of neighbouring points under the assumption that the surface geometry is locally smooth. Various spatial phase unwrapping algorithms have been developed, including those summarised by Ghiglia and Pritt (1998), Su and Chen (2004), and Zhao *et al* (2011), with some being more robust than others. Spatial phase unwrapping algorithms have been extensively used, yet the fundamental issues of all spatial phase unwrapping algorithms are: (i) the surface *smoothness* assumption limits their applicability to a single continuous surface without sharp changes—less than π between two adjacent pixels along at least one path; and (ii) the fundamental working mechanism (relative to one point) naturally cannot be used for absolute measurement since only *relative* unwrapped phase can be obtained.

Temporal phase unwrapping methods, in contrast, determine the *absolute* fringe order from additionally acquired information. Temporal phase unwrapping often refers to two- and multi-wavelength algorithms (Cheng and Wyant 1984, 1985, Towers *et al* 2003) that were developed for interferometry systems. Due to the flexibility of FPP techniques, temporal phase unwrapping algorithms are more diverse, including those using temporally acquired two- or multi-frequency phase-shifting patterns (Guan *et al* 2003, Li *et al* 1997), binary coded patterns (Hall-Holt and Rusinkiewicz 2001, Zhang *et al* 2012) and stair-phase encoded patterns (Wang and Zhang 2012, Zhou *et al* 2015). Temporal phase unwrapping methods typically unwrap each pixel independently and thus they fundamentally eliminate the surface smoothness constraints of spatial phase unwrapping. However, compared to spatial phase unwrapping, temporal phase unwrapping hinges on capturing additional information temporarily and, thus, slows down overall measurement speed and introduces more artefacts caused by motion.

5.4.2 Non-conventional phase unwrapping methods

Due to the rapidly reduced cost of hardware components, developing FPP systems with more hardware components has become less problematic. As such, novel *geometric-constraint-based phase unwrapping* algorithms have been developed. These algorithms introduce additional geometric constraints provided by adding hardware components (for example, a projector or camera) to determine fringe orders without capturing additional information temporarily (Bräuer-Burchardt *et al* 2013, Ishiyama *et al* 2007, Li *et al* 2013, Zhong *et al* 2013). As a result, they are suitable for high-speed applications. However, the field of view (FOV) of the overall measurement system is reduced because all target points have to be in the common visible areas of all hardware components. Furthermore, the complexity and cost of the system increase accordingly.

Without adding hardware components, An *et al* (2016) developed a method that uses the inherent geometric constraints of a standard camera-projector FPP system to virtually create a phase map on the projector at a virtual plane, with minimum depth. The artificially created phase map is called the minimum phase Φ_{mim} and the method is called the *minimum phase* method. The minimum phase Φ_{mim} is used to determine the fringe order per pixel independently. Figure 5.3 shows an example of performing absolute 3D shape measurement using this phase unwrapping method. This algorithm is robust and works well, however, it has a limited maximum depth range over which it can theoretically operate.

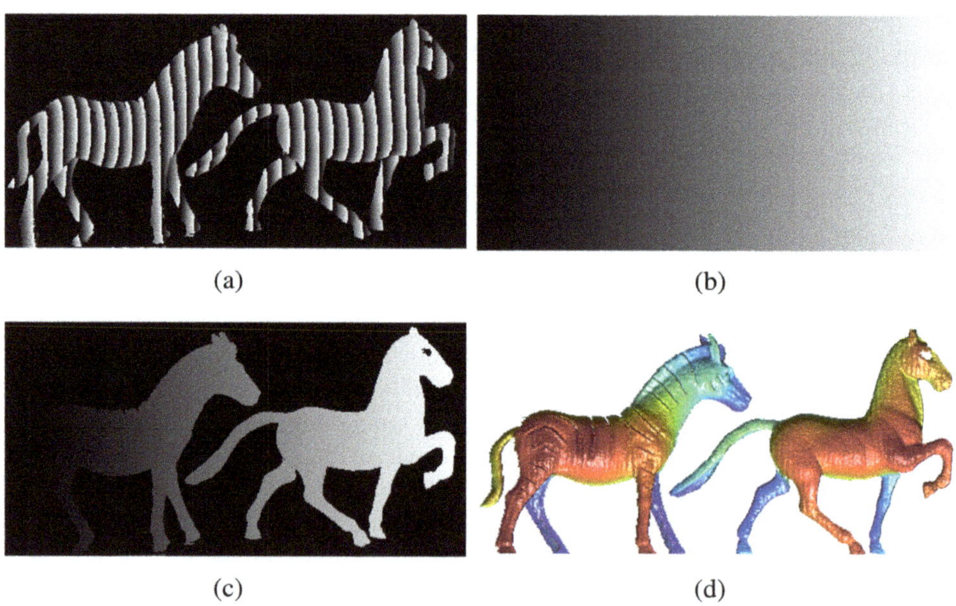

Figure 5.3. Examples of using a pre-knowledge-based phase unwrapping method for absolute 3D shape measurement: (a) wrapped phase map; (b) virtually created minimum phase map; (c) unwrapped phase map; and (d) 3D reconstruction using the unwrapped phase map. Adapted with permission from An *et al* (2016). Copyright 2016 The Optical Society.

Pre-knowledge-based phase unwrapping unwraps the absolute phase by referring to some pre-knowledge of the object, such as depth range or 3D geometry. When the computer-aided-design (CAD) models (Li *et al* 2017) or approximate 3D shape (An and Zhang 2019) of the target object is available, the absolute phase can be unwrapped by referring to prior geometry information combined with the minimum phase approach. These methods do not require additional projected patterns and are suitable for high-speed applications, but they are restricted to performing measurements with prior knowledge of the object.

Hybrid phase unwrapping methods have also been developed, including combining minimum phase based phase unwrapping with traditional temporal phase unwrapping (Hyun and Zhang 2016), phase-coding with pre-knowledge based phase unwrapping (Hyun and Zhang 2017), stereo-vision combined with phase constraints (Lohry *et al* 2014, Lohry and Zhang 2014), marker encoding with conventional spatial phase unwrapping (Budianto *et al* 2014, Cui *et al* 2012, Zhang and Yau 2006), along with others.

5.4.3 Machine-learning-based phase unwrapping methods

Recently, machine learning has been introduced to tackle the challenging phase unwrapping problem. Yin *et al* (2019) used machine learning to improve multi-frequency phase unwrapping accuracy and robustness. The input to their CNNs is two relative phase maps of different frequencies, and the output is the fringe order number. Experiments validated that the learning-based phase unwrapping method can obtain the absolute phase map accurately and reliably even when the phase-shifting step number is small and the phase noise is relatively large. Feng *et al* (2019) improved the method of Yin *et al* (2019) by using CNNs to predict the phase map from three single fringe patterns of different frequencies. Due to the reduction of projected patterns, this method is more applicable for high-speed applications. Other machine learning-based phase unwrapping algorithms have also been developed to analyse fringe patterns produced by other means (Spoorthi *et al* 2018, Wang *et al* 2019, Zhang *et al* 2019a, 2019b). These methods could also be beneficial to fringe projection. The main limitation of learning-based phase unwrapping methods is that they all require abundant data of accurate ground truths for training. Furthermore, the generalisability of such methods to scenarios different from the training data still need to be evaluated and validated.

5.5 High dynamic range techniques

One of the active research areas in the FPP field is to study how to accurately measure high-contrast or high dynamic range (HDR) surfaces, including surfaces that are partially specular. Similar to 2D imaging techniques under similar situations, various 3D measurement techniques have been developed to tackle this challenging problem. Basically, these HDR 3D measurement techniques endeavor to reduce the intensity in specular and/or saturation regions by adjusting the hardware settings (for example, changing the camera exposure and/or the projector's projection characteristics), such that the captured fringe patterns have an overall higher signal-to-noise ratio (SNR) than the image before making such adjustments; or by acquiring additional images captured or illuminated from different perspectives and selectively using the higher quality fringe images. State-of-the-art methods can be broadly classified into software, hardware or hybrid approaches (Feng *et al* 2018). The hardware approach to tackle this problem is primarily to use a polarisation filter in front of the system (Chen *et al* 2007, Yoshinori *et al* 2003), or to replace the standard camera with a camera having multiple polarisations (Salahieh *et al* 2014). Although effective, such a methods sacrifice the light throughput and can only partially reduce the influences of image glare. The software methods reconstruct 3D shape without altering the hardware and overcome the image contrast problem by projecting additional inverted fringe patterns (Jiang *et al* 2016), increasing the number of phase-shifted fringe images (Chen and Zhang 2016), properly timing the captured images (Suresh *et al* 2018), using different responses of the coloured camera pixels (Yin *et al* 2017, Zheng *et al* 2019), leveraging the high spectral response of the system (Wang *et al* 2018), and other methods (Cai *et al* 2016, Chen and Zhang 2016, Hu *et al* 2010, Qi *et al* 2018). All these methods are preferred when the hardware adjustment is neither practical nor desirable. However, their capabilities are rather limited.

There has been considerable effort to develop hybrid methods because they tend to be more powerful and robust than purely software or hardware adjustments. The hybrid method basically requires the change of hardware settings and the adjustment of software algorithms accordingly. HDR can be realised by artificially creating fringe patterns for 3D reconstruction from a sequence of images captured under different camera settings (for example, exposure time) (Feng *et al* 2014, Long *et al* 2015, Qi *et al* 2016, Rao and Da 2018, Song *et al* 2017, Zhang and Yau 2009, Zhang *et al* 2015, Zhang 2020, Zhong *et al* 2015) or with different projected images (Babaie *et al* 2015, Chen *et al* 2018a, 2018b, Jiang *et al* 2012, Lin *et al* 2016, Lin *et al* 2017, Qi *et al* 2018, Ri *et al* 2008, Sheng *et al* 2017, Waddington and Kofman 2010, Zhang 2018, Zhao *et al* 2014). Figure 5.4 shows one example of the rapid auto-exposure

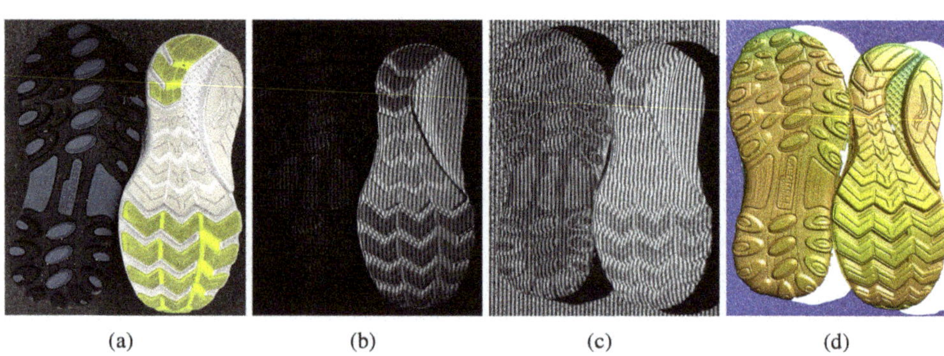

Figure 5.4. Example of measuring a high-contrast scene with an HDR algorithm (Zhang 2020): (a) photograph of the high contrast scene; (b) one of the fringe images captured with one of the exposure times; (c) one of the artificially created fringe images; and (d) 3D result. Reprinted from Zhang (2020), copyright 2020, with permission from Elsevier.

time algorithm developed by Zhang (2020), and the rapid auto-exposure time determination algorithm was further adapted to handle HDR situations. These results show that a high-contrast scene can be captured with fine detail fidelity. Despite their improvements over just software or hardware methods, the complexity of hybrid methods is increased, and it often takes more time for them to complete the measurement process.

5.6 Calibration

System calibration plays a key role in any metrology system. FPP systems are conventionally calibrated in two ways: adopting the reference plane approach that was developed for interferometry systems and geometrically calibrating both camera and projector with a standard pinhole camera model. It should be noted that almost all FPP system users refer to this process as calibration, but it is not calibration in the sense of that defined in chapter 1 as it does not always involve statements of uncertainty at each stage. However, the term is so ubiquitous that it will be used in the rest of this chapter. Lately, researchers have endeavored to develop methods to address issues related to lens defocusing or lens artefacts, as well as the use of telecentric lenses. This section covers the conventional system calibration methods as well as those recently developed.

5.6.1 Conventional methods

The *reference plane* based approach was previously used for FPP system calibration. In this method, an 'ideal' planar surface is measured as the reference plane; any measured geometry deviates relative to this reference plane. This calibration approach can work well if both the projector and the camera use telecentric lenses, or the system has a relatively small depth of field. However, such an approach can introduce large measurement errors if the system uses pinhole lenses or is required to measure an object with a large depth range. Furthermore, the reference plane measurement is required to be parallel to the projector–camera baseline to achieve high accuracy (Takeda and Mutoh 1983), which poses practical challenges for high-accuracy FPP system calibration.

The 'geometric calibration approach' typically models the camera and projector using a standard pinhole model (Zhang 2000). The pinhole model essentially models the linear transformation from the world coordinate system to the camera lens coordinate system; and the linear or nonlinear transformation from the lens coordinate system to the image plane. Camera calibration is well established and only requires capturing images of a flat calibration plane with some known feature points on the plane (for example, checkerboard, circle patterns), and then using existing open source software packages (for example, OpenCV camera calibration toolbox) to estimate the transformation parameters. Although numerous methods have been developed for structured light system calibration (Du and Wang 2007, Guo *et al* 2006, Hu *et al* 2003, Huang *et al* 2010, Mao *et al* 2007, Vo *et al* 2012, Zappa *et al* 2009), the most popular method is the one developed by Zhang and Huang (2006). For such a method, the projector can 'capture' images in a similar way to that of a camera (facilitated by the camera), and thus the projector calibration becomes the same as the camera calibration. Despite its popularity, the 'standard' pinhole model may not work well for high-accuracy 3D surface measurement because it cannot accurately model lens artefacts for standard, cost-effective lenses. Furthermore, these geometric calibration methods only work well if the projector and the camera are close to being in focus.

5.6.2 Recent developments

To address the lens defocusing issue, Li *et al* (2014) showed that a one-to-one mapping can be established even if the projector is out of focus. Therefore, instead of creating a projector image, the mapped feature points in a continuous space are directly used for projector calibration. Similarly, the camera might also be out of focus, and accurately calibrating a severely defocused camera is often difficult. Bell and Zhang (2016) developed a method to address the camera defocus problem. Bell *et al*'s method uses an LCD monitor to create active fringe patterns with encoded feature points in the phase domain. Since the phase information is accurately preserved, even if the camera is not focused, the encoded feature points can be accurately extracted for calibration, regardless of the degree of lens focusing.

Improvements and innovations to the general calibration method have also been made recently. For instance, Yin *et al* (2012) used a bundle adjustment strategy, Huang *et al* (2010) employed least-squares algorithms for system parameter estimation, An *et al* (2016) developed a method for large-scale system calibration and Vargas *et al* (2020) developed a hybrid method that further improves the calibration accuracy.

Among these methods, to our knowledge, the hybrid method (Vargas *et al* 2020) probably works the best. Figure 5.5 shows a comparison using the hybrid method (Vargas *et al* 2020) and the traditional method with camera and projector being modelled as a pinhole lens, with nonlinear distortion included. From the error map shown in figure 5.5(a), the irregular error distributions across the image cannot be easily modelled by the single mathematical function assumed by the pinhole model. From the error distributions of the planar surface measurement shown in figure 5.5(c), it can be noticed that i) the hybrid model produces errors following

(a) (b) (c)

Figure 5.5. Comparing the result of measuring an object with the hybrid calibration algorithm versus the traditional algorithm: (a) 3D result using standard calibration with a pinhole model; (b) 3D result using the hybrid model; and (c) planar error distributions. Adapted with permission from Vargas *et al* (2020). Copyright 2020 The Optical Society.

the normal distribution, while the traditional stereo-pinhole model does not; (ii) the planar error using the hybrid model does not spread as much as the stereo pinhole model ([−0.1 + 0.1] mm against [−0.2, +0.2] mm); and (iii) the standard deviation of the planar error is significantly reduced with the hybrid calibration approach (from 0.188 mm to 0.044 mm).

Even with recent developments, accurately calibrating an FPP system remains non-trivial. The mathematical modelling of a lens may have to be fundamentally changed to further improve calibration accuracy. For example, a more complex model, such as ray-tracing, may be necessary. Instead of representing the camera imaging system as a smooth function, the ray-tracing method considers each pixel ray path independently, and thus the local distortions caused by lens artefacts can be accurately represented. The challenge, however, is that there is no mature calibration approach to estimate the ray path for each pixel.

5.7 High-speed FPP realisation

To capture dynamic scenes, both fringe projection and image acquisition speeds must be high. Limited by hardware, conventional high-speed 3D imaging technologies typically use a static pattern projection to achieve limited accuracy and resolution, whilst in more recent work, researchers tend to use multiple fringe patterns by leveraging hardware advancements to achieve better accuracy and higher speed. This section discusses high-speed technologies that have been realised for FPP.

5.7.1 Conventional high-speed FPP methods

The single static fringe pattern projection methods can achieve the highest possible speeds (i.e. as fast as the camera captures images). The single fringe pattern can be a static monochromatic or colour fringe pattern (Huang *et al* 1999, Pan *et al* 2006, Wust and Capson 1991) projected by a projection unit and captured by an image acquisition unit. The former typically refers to FTP, which has been demonstrated successfully for rapid event capture (Takeda 2013, Su and Zhang 2010). The latter is more applicable for phase-shifting profilometry, to potentially achieve higher resolution and better capability (Zhang 2012). However, it may be challenging for a color-based method to measure colourful objects with high quality. It is interesting to note that it is not necessary for an FTP system to use a high-speed camera to capture rapidly changing scenes without motion introduced errors if a strobe light is used (Zhang *et al* 2005).

Since the object surface color texture can introduce problems if colour fringe patterns are used, rapidly changing multiple fringe patterns are necessary for high-speed and high-accuracy 3D shape measurement. The DLP projector has a unique projection mechanism that sequentially projects RGB channels at a video rate (60–120 Hz). After removing colour filters, three phase-shifted monochromatic fringe patterns were projected at a video rate. Taking advantage of this projection technology, Zhang and Huang (2006) have developed a real-time 3D shape measurement that achieved 40Hz, and later (Zhang and Yau 2006) developed a system that achieved 60 Hz.

5.7.2 Recent developments

Limited by the refresh rate of commercial video projectors, 3D shape measurement speeds have reached a bottleneck. Advanced hardware technologies offer new opportunities for higher speed 3D shape measurements. As discussed in section 5.2.2, speed breakthroughs were enabled by deploying the binary defocusing technique (Lei and Zhang 2009) using a DLP development kit that can switch binary images at tens of kilohertz (Xu and Zhang 2020). Researchers also developed high-speed mechanical projectors (Heist *et al* 2016, Hyun *et al* 2018) or LED arrays (Fujigaki *et al* 2013, Heist *et al* 2014) for high-speed 3D shape measurement and overcame some limitations of DLP technology. Furthermore, fibre optics was used to create phase-shifted interference fringe patterns at frame rates of gigahertz (Li *et al* 2014). Since advanced hardware technologies offer new opportunities for high-speed 3D shape measurement capabilities, the 3D shape measurement field is expected to continue to progress.

High-speed 3D shape measurement not only involves high-speed fringe pattern acquisition, but also 3D reconstruction and visualisation. Researchers have attempted to use massively parallel processors (Gao and Kemao 2012, Wang and Kemao 2018), such as grid computing, graphics processing units (GPUs) (Feng *et al* 2015, Karpinsky *et al* 2014, Liu *et al* 2010, Wang *et al* 2019, Zhang *et al* 2006), and field programmable gate arrays (FPGAs) (Wang *et al* 2016, Zhan *et al* 2017). Since other types of structured light 3D sensors have been embedded on mobile devices (for example, smart phones and tablets), the hardware platform developed for these sensors, such as advanced RISC machines (ARMs) and digital signal processing (DSP) platforms, may be leveraged to further advance the high-speed FPP capabilities.

Figure 5.6 shows some representative frames when measuring a live beating rabbit heart using a binary defocusing technique on the DLP platform. The projector projects binary structured patterns at 2000 Hz, and the camera, precisely synchronised with the projector, captures the images for 3D reconstruction. A three-

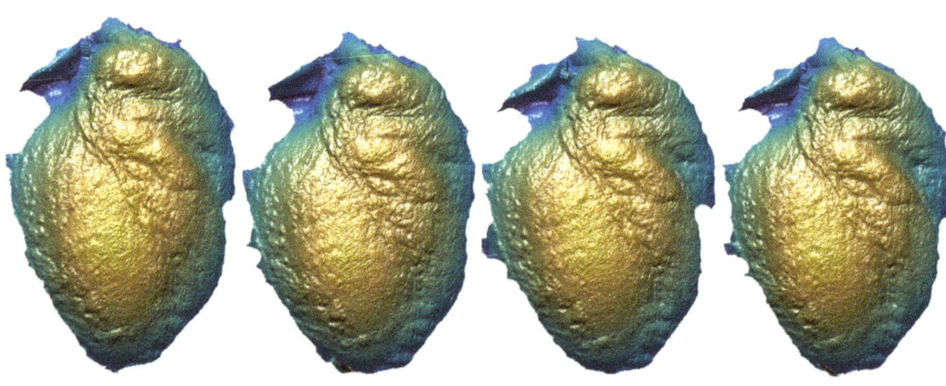

Figure 5.6. 3D representative frames of measuring a live beating rabbit heart. Reprinted from Zhang (2018), copyright 2018, with permission from Elsevier.

step phase-shifting algorithm was used to retrieve the phase, and grey coded binary patterns were used to determine fringe orders for temporal phase unwrapping.

5.8 Towards automation

Measurement data quality hinges on the quality of the acquired fringe images as well as the employed analysis software algorithms. For a given 3D reconstruction algorithm, how to acquire high-quality data becomes critical to achieving high-quality 3D shape measurement. Unfortunately, most FPP systems require skillful/trained personnel to acquire the best quality data. To make such a technology ubiquitous, system automation is one of the most important issues to address.

Unlike 2D imaging, where auto-exposure and HDR algorithms are relatively mature, 3D FPP still lacks algorithms to automatically determine the best exposure (s) for optimal quality data capture. The problems associated with FPP systems are more complex than those associated with 2D imaging systems because of the use of the projector. The problems become even more pronounced when a DLP projector is adopted for fringe projection. The fundamental working mechanism of a DLP projector limits the use of discrete exposure times if 8 bit grayscale fringe patterns are being sent to the projector for projection. As a result, optimal exposure times might not be practically available.

The binary defocusing technique, in contrast, allows the use of an arbitrary exposure time, making it possible to capture fringe patterns with the optimal exposure time. For example, Ekstrand and Zhang (2011) developed an algorithm to determine the optimal exposure time by capturing a sequence of images with varying exposure times. Although this approach works well, such a method is not practical because of its slow speed. As discussed earlier, various HDR algorithms have also been developed for FPP systems but they cannot determine proper exposures rapidly. To address this issue, Zhang (2020) developed a method that can determine the single global optimal exposure time by capturing image(s) with a single exposure time and further determine the HDR exposure times by capturing another image(s) with the optimal exposure time. This method shows promise for auto-exposure time control of FPP systems.

Yet another aspect of automation is auto-focus, which has largely been ignored in the FPP community because of its very challenging nature. As a result, state-of-the-art FPP systems are designed with a fixed focal depth range. Recently, researchers demonstrated the use of a newly developed electrically tunable lens (ETL) to tackle this challenging problem. ETL is electronically controllable and is repeatedly stable to a high degree. For example, Hu *et al* (2020) showed that the ETL could be modelled as a standard form of pinhole model except where all parameters are varying (for example, focal length, principle points), and those parameters were pre-calibrated as a function of the driving electrical signal of the ETL. However, the drawback of this approach is that the repeatability of ETL still cannot meet the standard required by high-accuracy metrology. Moreover, changing system orientation alters almost all calibration parameters, making it practically difficult to be implemented on portable systems. Zhong *et al* (2020) circumvented this problem by

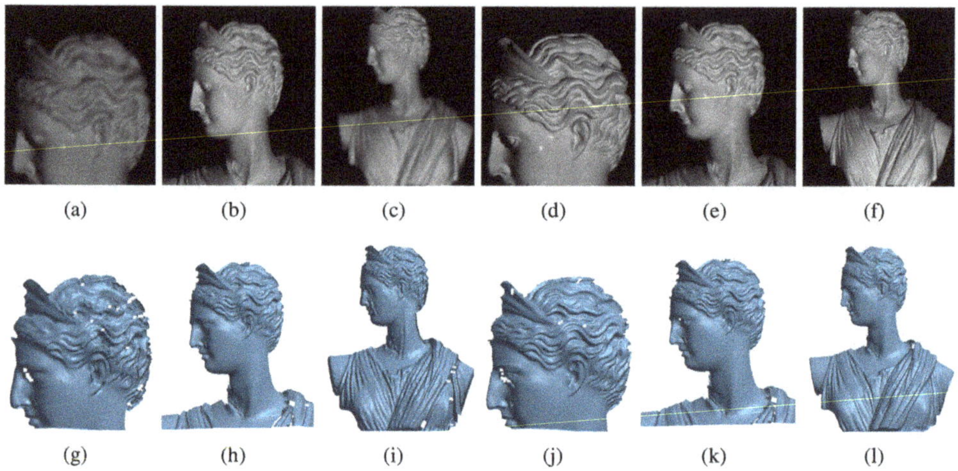

Figure 5.7. Example of measuring a 3D object using an auto-focusing method: (a) a fixed focal plane at approximately $z \approx 750$ mm when the object is at $z \approx 450$ mm; (b) a fixed focal plane at approximately $z \approx 750$ mm when the object is at $z \approx 750$ mm; (c) a fixed focal plane at approximately $z \approx 750$ mm when the object is at $z \approx 1050$ mm; (d)–(f) corresponding images for the use of the auto-focusing technique; and (g)–(l) corresponding 3D results. Adapted with permission from Hu *et al* (2020). Copyright 2020 The Optical Society.

adding a second projector. For such a system, 3D reconstruction was realised by matching points for these two calibrated projectors. The camera with an ETL only serves as the 'bridge' to establish the corresponding pairs. As such, the camera can freely adjust the focal plane to ensure that the object is focused without being calibrated. This approach could provide an option for auto-focus, albeit the complexity of the system is drastically increased and the interference between two projectors has to be resolved for high-speed applications.

Figure 5.7 shows the results obtained from the autofocusing technique (Hu *et al* 2020) when the object was at approximately $z \approx 450$ mm, $z \approx 750$ mm and $z \approx 1050$ mm away from the system. Figure 5.7(a)–(c) show the photographs of the object when the focal plane is fixed at $z \approx 750$ mm with a fixed driving current for ETL. Figures 5.7(d)–(f) show the photographs of the same object with the autofocusing technique when the driving current of ETL was adjusted properly. Figures 5.7(g)–(l) show the corresponding 3D reconstructions. Clearly, the autofocusing technique could result in focused 2D images and detailed 3D reconstructions for all three positions without a noticeable defocusing effect.

5.9 Towards integrated solutions

3D optical metrology systems are historically developed as an instrument, and leave application problems to be solved using either a one-size-fits-all approach or customisation with substantial costs. Such an approach may not always be the best way to solve practical problems. There has been a great deal of research recently to develop fully integrated solutions that can optimise the entire system for practical applications. In this section we present two solutions to demonstrate this concept.

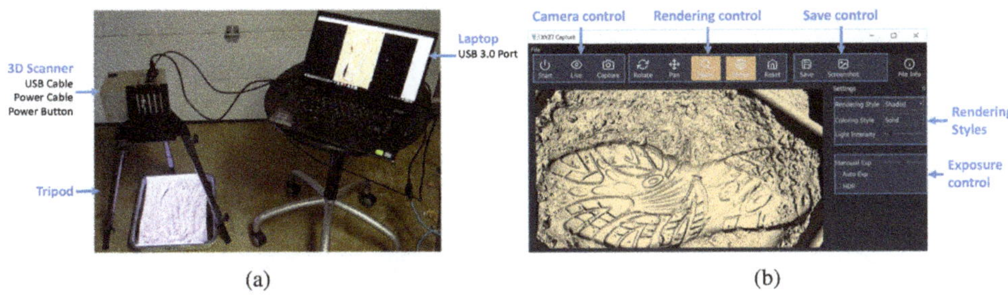

Figure 5.8. Portable 3D imaging system for crime scene documentation (Xu and Zhang 2020): (a) hardware system and (b) software graphical user interface (GUI). Reprinted from Xu and Zhang (2020), copyright 2020, with permission from Elsevier.

5.9.1 3D imaging system for crime scene evidence collection

Crime scene evidence collection plays a key role in law enforcement and national security. Conventional evidence collection approaches with high-resolution 2D photography face large challenges in courts due to the loss of accurate dimensional information, and the alternative casting method has been proved to be expensive with numerous associated issues. High-resolution 3D imaging technology could provide a means to document crime scenes. However, the state-of-the-art high-resolution 3D imaging technologies are not portable, affordable or simple to use. The desired system is a fully automated portable and affordable device that can be operated by anyone without training to capture desired high-quality evidence.

Working with a team of forensic scientists, engineers and practitioners, we spent more than three years developing the prototype system shown in figure 5.8(a). The system can be set up quickly with four steps for consistent high-quality 3D data capture: (i) plug in the power cord; (ii) connect the USB cable between the device and a laptop computer; (iii) toggle the power switch on; and (iv) start the software for data capture. The auto-exposure control algorithms (Zhang 2020) power the entire fully automated data acquisition process, allowing anyone to operate the system without training. The software graphical user interface (GUI), shown in figure 5.8(b) was designed based on the feedback from numerous practitioners. The GUI is intuitive and simple to use on both personal computers and touch screen tablets.

Figure 5.9 shows a representative example measurement. The 3D result shown in figure 5.9(d) was captured by a retired forensic examiner using the auto-exposure time control option provided by the system. As can be seen, the quality of data is high. This level of integration would not be possible without closely working with domain experts (i.e. those practitioners and scientists in crime scene evidence collection).

5.9.2 Robotic path planning

Accurate and efficient 3D reconstruction of the surrounding environment and target objects is critical to almost all robot applications, including navigation, path

Figure 5.9. Example of crime scene evidence documentation (Xu and Zhang 2020): (a) a high-resolution photograph of the shoe; (b) a high-resolution photograph of the tire impression in clay silt; (c) a high resolution photograph of the cast; and (d) 3D images of the impression. Reprinted from Xu and Zhang (2020), copyright 2020, with permission from Elsevier.

planning and drilling. For robot path planning, there are two kinds of methods: offline programming and online programming. Offline programming is usually performed by professional operators controlling the robot's movement to a sequence of key locations (Ang *et al* 2000). However, this process can be time-consuming and needs to be performed for each workpiece, even if the difference in geometry is slight. In contrast, online methods utilise CAD data to generate paths automatically, and are widely used in industrial applications. 3D surface measurement systems can be used to measure parts and software algorithms are used to analyse the difference between a CAD model and the actual workpiece through online scanning and registration (Rodrigues *et al* 2013). However, most of the state-of-the-art methods directly use point cloud data captured by a 3D sensor for analysis and robotic control (Rao *et al* 2017). Analysing point cloud data is complex and time consuming. Instead of using reconstructed point clouds, the phase map retrieved from an FPP system can be directly used to control robot movement (Rao *et al* 2020). By doing so, the 3D data processing problem becomes a 2D image processing problem, significantly improving the efficiency of the robotic system. With an FPP system, researchers demonstrated that the end-effector of an industrial robot can be

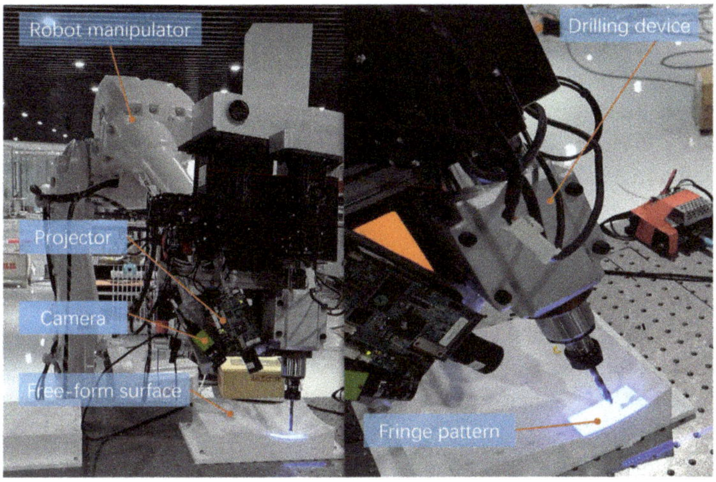

Figure 5.10. A robot drilling system facilitated with an FPP system: the robot end effector is controlled using the generated phase map directly without reconstructing the 3D point cloud for higher computational efficiency. Reprinted from Xu and Zhang (2020), copyright 2020, with permission from Elsevier.

adjusted to be perpendicular to a planar surface (Chen *et al* 2018), a cylindrical surface (Li *et al* 2019) or a freeform surface (Rao *et al* 2020).

Figure 5.10 shows a real application of an FPP system for high-accuracy robotic drilling control using the FPP system. Once again, the integrated solution could not have been developed without experts in robotics and experts in 3D optical metrology closely working together to create an integrated solution.

5.10 Summary

In summary, substantial advancements have been made recently in the field of fringe projection profilometry in all aspects from fringe generation to fringe analysis, calibration, as well as integrated solutions. However, there is still great need for further development to make such technologies more valuable in real applications. We envision that more integrated solutions will be developed through interdisciplinary collaborations to fully leverage the potential of system-level optimisation.

References

Abid A Z, Gdeisat M A, Burton D R, Lalor M J and Lilley F 2007 Spatial fringe pattern analysis using the two-dimensional continuous wavelet transform employing a cost function *Appl. Opt.* **46** 6120–6

Ajubi G A, Ayubi J A, Di Martino J M and Ferrari J A 2011 Three-dimensional profiling with binary fringes using phase-shifting interferometry algorithms *Appl. Opt.* **50** 147–54

An Y, Bell T, Li B, Xu J and Zhang S 2016 Novel method for large range structured light system calibration *Appl. Opt.* **55** 9563–72

An Y, Liu Z and Zhang S 2016 Evaluation of pixel-wise geometricconstraint-based phase unwrapping method for low signal-to-noise-ratio (SNR) phase *Adv. Opt. Technol.* **5** 423–32

An Y, Hyun J-S and Zhang S 2016 Pixel-wise absolute phase unwrapping using geometric constraints of structured light system *Opt. Express* **24** 18445–59

An Y and Zhang S 2019 Pixel-by-pixel absolute phase retrieval assisted by an additional three-dimensional scanner *Appl. Opt.* **58** 2033–41

Anderson J A and Porter R W 1929 Ronchi's method of optical testing *Astrophys. J.* **70** 175–81

Ang M H, Wei L and Yong L S 2000 An industrial application of control of dynamic behavior of robots-a walk-through programmed welding robot *Proc. 2000 ICRA. Millennium Conf. Int. Conf. on Robotics and Automation. Symposia Proc.* vol 3 (Piscataway, NJ: IEEE) pp 2352–7

Ayubi G A, Ayubi J A, Di Martino J M and Ferrari J A 2010 Pulse-width modulation in defocused 3-D fringe projection *Opt. Lett.* **35** 3682–4

Babaie G, Abolbashari M and Farahi F 2015 Dynamics range enhancement in digital fringe projection technique *Precis. Eng.* **39** 243–51

Bell T and Zhang S 2016 Method for out-of-focus camera calibration *Appl. Opt.* **55** 2346–52

Brahm A, Rößler C, Dietrich P, Heist S, Kühmstedt P and Notni G 2016 Non-destructive 3D shape measurement of transparent and black objects with thermal fringes *Proc. SPIE* **9868** 98680C

Budianto B, Lun P K D and Hsung T-C 2014 Marker encoded fringe projection profilometry for efficient 3D model acquisition *Appl. Opt.* **53** 7442–53

Bräuer-Burchardt C, Kühmstedt P and Notni G 2013 Phase unwrapping using geometric constraints for high-speed fringe projection based 3D measurements *Modeling Aspects in Optical Metrology IV* vol 8789 (Bellingham, WA: International Society for Optics and Photonics) 878906

Cai Z, Liu X, Peng X, Yin Y, Li A, Wu J and Gao B Z 2016 Structured light field 3D imaging *Opt. Express* **24** 20324–34

Chen W, Su X, Cao Y P, Zhang Q C and Xiang L Q 2005 Method for eliminating zero spectrum in Fourier transform profilometry *Opt. Lasers Eng.* **43** 1267–76

Chen R, Han S, Xu J and Su H 2019 Point-based multi-view stereo network arXiv: 1908.04422

Chen T, Lensch H P A, Fuchs C and Seidel H-P 2007 Polarization and phase-shifting for 3D scanning of translucent objects *2007 IEEE Conf. on Computer Vision and Pattern Recognition* (Piscataway, NJ: IEEE) pp 1–8

Chen B and Zhang S 2016 High-quality 3D shape measurement using saturated fringe patterns *Opt. Laser Eng.* **87** 83–9

Chen C, Gao N, Wang X and Zhang Z 2018a Adaptive pixel-to-pixel projection intensity adjustment for measuring a shiny surface using orthogonal color fringe pattern projection *Meas. Sci. Technol.* **29** 055203

Chen C, Gao N, Wang X and Zhang Z 2018b Adaptive projection intensity adjustment for avoiding saturation in three-dimensional shape measurement *Meas. Sci. Technol.* **410** 694–702

Chen R, Wang G, Zhao J, Xu J and Chen K 2018 Fringe pattern based plane-to-plane visual servoing for robotic spray path planning *IEEE/ASME Trans. Mechatron.* **23** 1083–91

Cheng Y-Y and Wyant J C 1984 Two-wavelength phase shifting interferometry *Appl. Opt.* **23** 4539–43

Cheng Y-Y and Wyant J C 1985 Multiple-wavelength phase shifting interferometry *Appl. Opt.* **24** 804–7

Cui H, Liao W, Dai N and Cheng X 2012 A flexible phase-shifting method with absolute phase marker retrieval *Measurement* **45** 101–8

Dai J, Li B and Zhang S 2014 High-quality fringe patterns generation using binary pattern optimization through symmetry and periodicity *Opt. Laser Eng.* **52** 195–200

Du H and Wang Z 2007 Three-dimensional shape measurement with an arbitrarily arranged fringe projection profilometry system *Opt. Lett.* **32** 2438–40

Ekstrand L and Zhang S 2011 Auto-exposure for three-dimensional shape measurement with a digital-light-processing projector *Opt. Eng.* **50** 123603

Feng S, Chen Q, Gu G, Tao T, Zhang L, Hu Y, Yin W and Zuo C 2019 Fringe pattern analysis using deep learning *Adv. Photon* **1** 025001

Feng S, Zuo C, Yin W, Gu G and Chen Q 2019 Micro deep learning profilometry for high-speed 3D surface imaging *Opt. Lasers Eng.* **121** 416–27

Feng S, Zhang L, Zuo C, Tao T and Gu G 2018 High dynamic range 3D measurements with fringe projection profilometry: a review *Meas. Sci. Technol.* **29** 122001

Feng S, Zhang Y, Chen Q, Zuo C, Li R and Shen G 2014 General solution for high dynamic range three-dimensional shape measurement using the fringe projection technique *Opt. Laser Eng.* **59** 56–71

Feng S, Chen Q and Zuo C 2015 Graphics processing unit-assisted real-time three-dimensional measurement using speckle-embedded fringe *Appl. Opt.* **54** 6865–73

Fujigaki M, Oura Y, Asai D and Murata Y 2013 High-speed height measurement by a light-source-stepping method using a linear led array *Opt. Express* **21** 23169–80

Gao W and Kemao Q 2012 Parallel computing in experimental mechanics and optical measurement: a review *Opt. Lasers Eng.* **50** 608–17

Gdeisat M A, Burton D R and Lalor M J 2006 Spatial carrier fringe pattern demodulation by use of a two-dimensional continuous wavelet transform *Appl. Opt.* **45** 8722–32

Gdeisat M A, Abid A, Burton D R, Lalor M J, Lilley F, Moore C and Qudeisat M 2009 Spatial and temporal carrier fringe pattern demodulation using the one-dimensional continuous wavelet transform: recent progress, challenges, and suggested developments *Opt. Lasers Eng.* **47** 1348–61

Gdeisat M, Burton D, Lilley F and Arevalillo-Herráez M 2016 Fast fringe pattern phase demodulation using FIR Hilbert transformers *Opt. Commun.* **359** 200–6

Ghiglia D C and Pritt M D (ed) 1998 *Two-Dimensional Phase Unwrapping: Theory, Algorithms, and Software* (New York: Wiley)

Girshick R 2015 *Fast R-CNN Proc. of the IEEE Int. Conf. on Computer Vision* pp 1440–8

Gong Y and Zhang S 2010 Ultrafast 3-D shape measurement with an off-the-shelf DLP projector *Opt. Express* **18** 19743–54

Guan C, Hassebrook L G and Lau D L 2003 Composite structured light pattern for three-dimensional video *Opt. Express* **11** 406–17

Guo H, Chen M and Zheng P 2006 Least-squares fitting of carrier phase distribution by using a rational function in fringe projection profilometry *Opt. Lett.* **31** 3588–90

Guo H and Huang P 2008 3-D shape measurement by use of a modified Fourier transform method *Proc. SPIE* **7066** 70660E

Hall-Holt O and Rusinkiewicz S 2001 Stripe boundary codes for real-time structured-light range scanning of moving objects *Proc. Eighth Int. Conf. on Computer Vision. ICCV 2001* vol 2 (Piscataway, NJ: IEEE) pp 359–66

Harihara P 2006 *Basics of Interferometry* (New York: Academic)

Heist S, Mann A, Kühmstedt P, Schreiber P and Notni G 2014 Array projection of aperiodic sinusoidal fringes for high-speed three-dimensional shape measurement *Opt. Eng.* **53** 112208

Heist S, Lutzke P, Schmidt I, Dietrich P, Kühmstedt P, Tünnermann A and Notni G 2016 High-speed three-dimensional shape measurement using gobo projection *Opt. Laser Eng.* **87** 90–6

Hornbeck L J 1997 Digital light processing for high-brightness high-resolution applications *Proc. SPIE* **3013** 27–40

Hu Q, Huang P S, Fu Q and Chiang F-P 2003 Calibration of a three-dimensional shape measurement system *Opt. Eng.* **42** 487–93

Hu E, He Y and Chen Y 2010 Study on a novel phase-recovering algorithm for partial intensity saturation in digital projection grating phase-shifting profilometry *Optik* **121** 23–8

Hu X, Wang G, Hyun J-S, Zhang Y, Yang H and Zhang S 2020 Autofocusing method for high-resolution three-dimensional profilometry *Opt. Lett.* **45** 375–8

Huang P S, Hu Q, Jin F and Chiang F-P 1999 Color-encoded digital fringe projection technique for high-speed 3-D surface contouring *Opt. Eng.* **38** 1065–71

Huang L, Chua P S K and Asundi A 2010 Least-squares calibration method for fringe projection profilometry considering camera lens distortion *Appl. Opt.* **49** 1539–48

Hyun J-S, Chiu G T-C and Zhang S 2018 High-speed and high-accuracy 3D surface measurement using a mechanical projector *Opt. Express* **26** 1474–87

Hyun J-S and Zhang S 2020 High-speed three-dimensional absolute shape measurement with three binary patterns *Opt. Eng.* **59** 024104

Hyun J-S and Zhang S 2016 Enhanced two-frequency phase-shifting method *Appl. Opt.* **55** 4395–401

Hyun J and Zhang S 2017 Superfast 3D absolute shape measurement using five fringe patterns *Opt. Laser Eng.* **90** 217–24

Ishiyama R, Sakamoto S, Tajima J, Okatani T and Deguchi K 2007 Absolute phase measurements using geometric constraints between multiple cameras and projectors *Appl. Opt.* **46** 3528–38

Jiang C, Bell T and Zhang S 2016 High dynamic range real-time 3D shape measurement *Opt. Express* **24** 7337–46

Jiang H, Zhao H and Li X 2012 High dynamic range fringe acquisition: a novel 3-D scanning technique for high-reflective surfaces *Opt. Laser Eng.* **50** 1484–93

Karpinsky N, Hoke M, Chen V and Zhang S 2014 High-resolution, real-time three-dimensional shape measurement on graphics processing unit *Opt. Eng.* **53** 024105

Kemao Q 2004 Windowed Fourier transform for fringe pattern analysis *Appl. Opt.* **43** 2695–702

Kemao Q, Wang H and Gao W 2008 Windowed Fourier transform for fringe pattern analysis: theoretical analyses *Appl. Opt.* **47** 5408–19

Kemao Q 2015 Applications of windowed Fourier fringe analysis in optical measurement: a review *Opt. Lasers Eng.* **66** 67–73

Krizhevsky A, Sutskever I and Hinton G E 2012 ImageNet classification with deep convolutional neural networks *NIPS'12: Proc. of the 25th Int. Conf. on Neural Information Processing Systems* vol 1 pp 1097–105

Lei S and Zhang S 2009 Flexible 3-D shape measurement using projector defocusing *Opt. Lett.* **34** 3080–2

Li B, Ou P and Zhang S 2014 High-speed 3D shape measurement with fiber interference *Proc. SPIE* **9203** 920310

Li B and Zhang S 2018 Novel method for measuring dense 3D strain map of robotic flapping wings *Meas. Sci. Technol.* **29** 045402

Li J, Chen Z, Rao G and Xu J 2019 Structured light-based visual servoing for robotic pipe welding pose optimization *IEEE Access* **7** 138327–40

Li X-X and Zhang Z-J 2017 High-quality fringe pattern generation based on binary pattern optimization with projector defocusing *J. Opt. Technol.* **84** 22–8

Li J, Su X and Guo L 1990 Improved Fourier transform profilometry for the automatic measurement of three-dimensional object shapes *Opt. Eng.* **29** 1439–45

Li J-L, Su H-J and Su X-Y 1997 Two-frequency grating used in phase-measuring profilometry *Appl. Opt.* **36** 277–80

Li Z, Zhong K, Li Y F, Zhou X and Shi Y 2013 Multiview phase shifting: a full-resolution and high-speed 3D measurement framework for arbitrary shape dynamic objects *Opt. Lett.* **38** 1389–91

Li B, Bell T and Zhang S 2017 Computer-aided-design (CAD) model-assisted absolute three-dimensional shape measurement *Appl. Opt.* **56** 6770–6

Li B, Karpinsky N and Zhang S 2014 Novel calibration method for structured light system with an out-of-focus projector *Appl. Opt.* **53** 3415–26

Lin H, Gao J, Mei Q, He Y, Liu J and Wang X 2017 Three-dimensional shape measurement technique for shiny surfaces by adaptive pixel-wise projection intensity adjustment *Opt. Lasers Eng.* **91** 206–15

Lin H, Gao J, Mei Q, He Y, Liu J and Wang X 2016 Adaptive digital fringe projection technique for high dynamic range three-dimensional shape measurement *Opt. Express* **24** 7703–18

Liu Z, Zibley P C and Zhang S 2018 Motion-induced error compensation for phase shifting profilometry *Opt. Express* **26** 12632–7

Liu K, Wang Y, Lau D L, Hao Q and Hassebrook L G 2010 Dual-frequency pattern scheme for high-speed 3-D shape measurement *Opt. Express* **18** 5229–44

Lohry W and Zhang S 2013 Genetic method to optimize binary dithering technique for high-quality fringe generation *Opt. Lett.* **38** 540–2

Lohry W, Chen V and Zhang S 2014 Absolute three-dimensional shape measurement using coded fringe patterns without phase unwrapping or projector calibration *Opt. Express* **22** 1287–301

Lohry W and Zhang S 2014 High-speed absolute three-dimensional shape measurement using three binary dithered patterns *Opt. Express* **22** 26752–62

Long Y, Wang S, Wu W and Liu K 2015 Accurate identification of saturated pixels for high dynamic range measurement *Opt. Eng.* **54** 043106

Lu F, Wu C and Yang J 2019 Optimized dithering technique for three-dimensional shape measurement with projector defocusing *Opt. Commun.* **430** 246–55

Luo F, Chen W and Su X 2016 Eliminating zero spectra in Fourier transform profilometry by application of Hilbert transform *Opt. Commun.* **365** 76–85

Mao X, Chen W and Su X 2007 Improved Fourier-transform profilometry *Appl. Opt.* **46** 664–8

Pan J, Huang P S and Chiang F-P 2006 Color phase-shifting technique for three-dimensional shape measurement *Opt. Eng.* **45** 013602

Qi Z, Wang Z, Huang J, Xue Q and Gao J 2016 Improving the quality of stripes in structured-light three-dimensional profile measurement *Opt. Eng.* **56** 031208

Qi Z, Wang Z, Huang J, Xing C and Gao J 2018 Error of image saturation in the structured-light method *Appl. Opt.* **57** A181–8

Qi Z, Wang Z, Huang J, Xing C and Gao J 2018 Highlight removal based on the regional-projection fringe projection method *Opt. Eng.* **57** 041404

Qian K 2010 Comparison of Fourier transform, windowed Fourier transform, and wavelet transform methods for phase extraction from a single fringe pattern in fringe projection profilometry *Opt. Laser Eng.* **48** 141–8

Rao G, Wang G, Yang X, Xu J and Chen K 2017 Normal direction measurement and optimization with a dense three-dimensional point cloud in robotic drilling *IEEE/ASME Trans. Mechatron.* **23** 986–96

Rao L and Da F 2018 High dynamic range 3D shape determination based on automatic exposure selection *J. Vis. Commun. Image Represent.* **50** 217–26

Rao G, Yang X, Yu H, Chen K and Jing X 2020 Fringe-projection-based normal direction measurement and adjustment for robotic drilling *IEEE Trans. Ind. Electron.* **67** 9560–70

Ren S, He K, Girshick R and Sun J 2015 Faster R-CNN: towards real-time object detection with region proposal networks *IEEE Trans. Pattern Anal. Mach. Intell.* **39** 1137–49

Ri S, Fujigaki M and Morimoto Y 2008 Intensity range extension method for three-dimensional shape measurement in phase-measuring profilometry using a digital micromirror device camera *Appl. Opt.* **47** 5400–7

Rodrigues M, Kormann M, Schuhler C and Tomek P 2013 Robot trajectory planning using OLP and structured light 3D machine vision *Int. Symp. on Visual Computing* (Berlin: Springer) pp 244–53

Rowe S H and Welford W T 1967 Surface topography of non-optical surfaces by projected interference fringes *Nature* **216** 786–7

Rusinkiewicz S, Hall-Holt O and Levoy M 2002 Real-time 3D model acquisition *ACM Trans. Graph.* **21** 438–46

Salvi J, Fernandez S, Pribanic T and Llado X 2010 A state of the art in structured light patterns for surface profilometry *Pattern Recogn.* **43** 2666–80

Salahieh B, Chen Z, Rodriguez J J and Liang R 2014 Multi-polarization fringe projection imaging for high dynamic range objects *Opt. Express* **22** 10064–71

Sheng H, Xu J and Zhang S 2017 Dynamic projection theory for fringe projection profilometry *Appl. Opt.* **56** 8452–60

Shi J, Zhu X, Wang H, Song L and Guo Q 2019 Label enhanced and patch based deep learning for phase retrieval from single frame fringe pattern in fringe projection 3D measurement *Opt. Express* **27** 28929–43

Silva A, Flores J L, Munoz A, Ayubi G A and Ferrari J A 2017 Three-dimensional shape profiling by out-of-focus projection of colored pulse width modulation fringe patterns *Appl. Opt.* **56** 5198–203

Simonyan K and Zisserman A 2014 Very deep convolutional networks for large-scale image recognition arXiv: 1409.1556

Song Z, Jiang H, Lin H and Tang S 2017 A high dynamic range structured light means for the 3D measurement of specular surface *Opt. Lasers Eng.* **95** 8–16

Spoorthi G E, Gorthi S and Gorthi R K S 2018 PhaseNet: a deep convolutional neural network for two-dimensional phase unwrapping *IEEE Signal Process. Lett.* **26** 54–8

Su X Y, Zhou W S, Von Bally G and Vukicevic D 1992 Automated phase-measuring profilometry using defocused projection of a Ronchi grating *Opt. Commun.* **94** 561–73

Su X and Chen W 2004 Reliability-guided phase unwrapping algorithm: a review *Opt. Laser Eng.* **42** 245–61

Su X and Zhang Q 2010 Dynamic 3-D shape measurement method: a review *Opt. Laser Eng.* **48** 191–204

Sun J, Zuo C, Feng S, Yu S, Zhang Y and Chen Q 2015 Improved intensity-optimized dithering technique for 3D shape measurement *Opt. Laser Eng.* **66** 158–64

Suresh V, Wang Y and Li B 2018 High-dynamic-range 3D shape measurement utilizing the transitioning state of digital micromirror device *Opt. Laser Eng.* **107** 176–81

Takeda M and Mutoh K 1983 Fourier transform profilometry for the automatic measurement of 3-D object shapes *Appl. Opt.* **22** 3977–82

Takeda M 2013 Fourier fringe analysis and its applications to metrology of extreme physical phenomena: a review *Appl. Opt.* **52** 20–9

Towers D P, Jones J D C and Towers C E 2003 Optimum frequency selection in multi-frequency interferometry *Opt. Lett.* **28** 1–3

Van der Jeught S and Dirckx J J J 2019 Deep neural networks for single shot structured light profilometry *Opt. Express* **27** 17091–101

Vargas R, Marrugo A G, Zhang S and Romero J A 2020 Hybrid calibration procedure for fringe projection profilometry based on stereo-vision and polynomial fitting *Appl. Opt.* **59** D163–9

Vo M, Wang Z, Pan B and Pan T 2012 Hyper-accurate flexible calibration technique for fringe-projection-based three-dimensional imaging *Opt. Express* **20** 16926–41

Waddington C and Kofman J 2010 Analysis of measurement sensitivity to illuminance and fringe-pattern gray levels for fringe-pattern projection adaptive to ambient lighting *Opt. Laser Eng.* **48** 251–6

Wang Y and Zhang S 2010 Optimal pulse width modulation for sinusoidal fringe generation with projector defocusing *Opt. Lett.* **35** 4121–3

Wang Y and Zhang S 2012 Novel phase-coding method for absolute phase retrieval *Opt. Lett.* **37** 2067–9

Wang Y and Zhang S 2012 Three-dimensional shape measurement with binary dithered patterns *Appl. Opt.* **51** 6631–6

Wang J, Xiong Z, Wang Z, Zhang Y and Wu F 2016 FPGA design and implementation of kinetic-like depth sensing *IEEE Trans. Circuits Syst. Video Technol.* **26** 1175–86

Wang Y, Jiang C and Zhang S 2017 Double-pattern triangular pulse width modulation technique for high-accuracy high-speed 3D shape measurement *Opt. Express* **25** 30177–88

Wang Y, Liu Z, Jiang C and Zhang S 2018 Motion induced phase error reduction using a Hilbert transform *Opt. Express* **26** 34224–35

Wang Y, Zhang J and Luo B 2018 High dynamic range 3D measurement based on spectral modulation and hyperspectral imaging *Opt. Express* **26** 34442–50

Wang T and Kemao Q 2018 Parallel computing in experimental mechanics and optical measurement: a review (II) *Opt. Lasers Eng.* **104** 181–91

Wang K, Li Y, Kemao Q, Di J and Zhao J 2019 One-step robust deep learning phase unwrapping *Opt. Express* **27** 15100–15

Wang H, Zeng H, Chen P, Liang R and Jiang L 2019 Fast single fringe-pattern processing with graphics processing unit *Appl. Opt.* **58** 6854–64

Wust C and Capson D W 1991 Surface profile measurement using color fringe projection *Mach. Vis. Appl.* **4** 193–203

Xiao Y and Li Y 2017 High-quality binary fringe generation via joint optimization on intensity and phase *Opt. Laser Eng.* **97** 19–26

Xu Z-X, Chan Y-H and Lun D P K 2017 High-quality octa-level fringe pattern generation for improving the noise characteristics of measured depth maps *Opt. Lasers Eng.* **98** 99–106

Xu J and Zhang S 2020 Status, challenges, and future perspectives of fringe projection profilometry *Opt. Laser Eng.* in press https://doi.org/10.1016/j.optlaseng.2020.106193

Yao Y, Luo Z, Li S, Fang T and Quan L 2018 MVSNet: depth inference for unstructured multi-view stereo *Proc. of the European Conf. on Computer Vision (ECCV)* pp 767–83

Yin Y, Peng X, Li A, Liu X and Gao B Z 2012 Calibration of fringe projection profilometry with bundle adjustment strategy *Opt. Lett.* **37** 542–4

Yin Y, Cai Z, Jiang H, Meng X, Xi J and Peng X 2017 High dynamic range imaging for fringe projection profilometry with single-shot raw data of the color camera *Opt. Lasers Eng.* **89** 138–44

Yin W, Chen Q, Feng S, Tao T, Huang L, Trusiak M, Asundi A and Zuo C 2019 Temporal phase unwrapping using deep learning *Sci. Rep.* **9** 20175

Yoshinori Y, Hiroyuki M, Osamu N and Tetsuo I 2003 Shape measurement of glossy objects by range finder with polarization optical system *Gazo Denshi Gakkai Kenkyukai Koen Yoko* **200** 43–50 (in Japanese)

Zappa E *et al* 2009 Fourier-transform profilometry calibration based on an exhaustive geometric model of the system *Opt. Laser Eng.* **47** 754–67

Zhan G, Tang H, Zhong K, Li Z, Shi Y and Wang C 2017 High-speed FPGA-based phase measuring profilometry architecture *Opt. Express* **25** 10553–64

Zhang Z 2000 A flexible new technique for camera calibration *IEEE Trans. Pattern Anal. Mach. Intell.* **22** 1330–4

Zhang S 2010 Recent progresses on real-time 3-D shape measurement using digital fringe projection techniques *Opt. Laser Eng.* **48** 149–58

Zhang Z H 2012 Review of single-shot 3D shape measurement by phase calculation-based fringe projection techniques *Opt. Laser Eng.* **50** 1097–106

Zhang S 2015 Comparative study on passive and active projector nonlinear gamma calibration *Appl. Opt.* **54** 3834–41

Zhang S 2018 High-speed 3D shape measurement with structured light methods: a review *Opt. Laser Eng.* **106** 119–31

Zhang S 2018 Absolute phase retrieval methods for digital fringe projection profilometry: a review *Opt. Laser Eng.* **107** 28–37

Zhang S 2020 Rapid and automatic optimal exposure control for digital fringe projection technique *Opt. Laser Eng.* **128** 106029

Zhang Q, Su X, Cao Y, Li Y, Xiang L and Chen W 2005 Optical 3-D shape and deformation measurement of rotating blades using stroboscopic structured illumination *Opt. Eng.* **44** 113601

Zhang S and Huang P S 2006 High-resolution real-time three-dimensional shape measurement *Opt. Eng.* **45** 123601

Zhang S and Yau S-T 2006 High-resolution, real-time 3-D absolute coordinate measurement based on a phase-shifting method *Opt. Express* **14** 2644–9

Zhang S and Huang P S 2006 Novel method for structured light system calibration *Opt. Eng.* **45** 083601

Zhang S, Royer D and Yau S-T 2006 GPU-assisted high-resolution, real-time 3-D shape measurement *Opt. Express* **14** 9120–9

Zhang S and Yau S-T 2007 High-speed three-dimensional shape measurement system using a modified two-plus-one phase-shifting algorithm *Opt. Eng.* **46** 113603

Zhang S and Yau S-T 2009 High dynamic range scanning technique *Opt. Eng.* **48** 033604

Zhang S, van der Weide D and Oliver J 2010 Superfast phase-shifting method for 3-D shape measurement *Opt. Express* **18** 9684–9

Zhang Q, Su X, Xiang L and Sun X 2012 3-D shape measurement based on complementary gray-code light *Opt. Lasers Eng.* **50** 574–9

Zhang B, Ouyang Y and Zhang S 2015 High dynamic range saturation intelligence avoidance for three-dimensional shape measurement *15th IACM Int. Symp. on Cluster, Cloud and Grid Computing* (Piscataway, NJ: IEEE) pp 981–90

Zhang S 2018 High-speed 3D shape measurement with structured light methods: a review *Opt. Lasers Eng.* **106** 119–31

Zhang J, Tian X, Shao J, Luo H and Liang R 2019a Phase unwrapping in optical metrology via denoised and convolutional segmentation networks *Opt. Express* **27** 14903–12

Zhang T, Jiang S, Zhao Z, Dixit K, Zhou X, Hou J, Zhang Y and Yan C 2019b Rapid and robust two-dimensional phase unwrapping via deep learning *Opt. Express* **27** 23173–85

Zhao M, Huang L, Zhang Q, Su X, Asundi A and Kemao Q 2011 Quality-guided phase unwrapping technique: comparison of quality maps and guiding strategies *Appl. Opt.* **50** 6214–24

Zhao H, Liang X, Diao X and Jiang H 2014 Rapid *in-situ* 3D measurement of shiny object based on fast and high dynamic range digital fringe projector *Opt. Laser Eng.* **54** 170–4

Zheng Y, Wang Y, Suresh V and Li B 2019 Real-time high-dynamic-range fringe acquisition for 3D shape measurement with a RGB camera *Meas. Sci. Tech.* **30** 075202

Zhong J and Weng J 2004 Spatial carrier-fringe pattern analysis by means of wavelet transform: wavelet transform profilometry *Appl. Opt.* **43** 4993–8

Zhong J and Zeng H 2007 Multiscale windowed Fourier transform for phase extraction of fringe patterns *Appl. Opt.* **46** 2670–5

Zhong K, Li Z, Shi Y, Wang C and Lei Y 2013 Fast phase measurement profilometry for arbitrary shape objects without phase unwrapping *Opt. Lasers Eng.* **51** 1213–22

Zhong K, Li Z, Zhou X, Li Y, Shi Y and Wang C 2015 Enhanced phase measurement profilometry for industrial 3D inspection automation *Int. J. Adv. Manuf. Technol.* **76** 1563–74

Zhong M, Hu X, Chen F, Xiao C, Peng D and Zhang S 2020 Autofocusing method for digital fringe projection system with dual projectors *Opt. Express* **28** 12609–20

Zhou C, Liu T, Si S, Xu J, Liu Y and Lei Z 2015 An improved stair phase encoding method for absolute phase retrieval *Opt. Lasers Eng.* **66** 269–78

Zhu J, Zhou P, Su X and You Z 2016 Accurate and fast 3D surface measurement with temporal–spatial binary encoding structured illumination *Opt. Express* **24** 28549–60

Zuo C, Chen Q, Feng S, Feng F, Gu G and Sui X 2012 Optimized pulse width modulation pattern strategy for three-dimensional profilometry with projector defocusing *Appl. Opt.* **51** 4477–90

Zuo C, Chen Q, Gu G, Feng S, Feng F, Li R and Shen G 2013 High-speed three-dimensional shape measurement for dynamic scenes using bi-frequency tripolar pulse-width-modulation fringe projection *Opt. Laser Eng.* **51** 953–60

Zuo C, Tao T, Feng S, Huang L, Asundi A and Chen Q 2018 Micro Fourier transform profilometry (μFTP): 3D shape measurement at 10,000 frames per second *Opt. Lasers Eng.* **102** 70–91

Chapter 6

Machine learning approaches

Joe Eastwood, Danny Sims-Waterhouse and Samanta Piano

Many of the form and coordinate metrology techniques discussed in this book, while effective, often rely on slow, user dependent procedures. The advent of high-performance parallel computing has allowed the field of machine learning to explode in popularity across many applications, including computer vision and metrology. This chapter serves as a broad introduction to machine learning approaches, then summarises recent work which has applications for the fields of form and coordinate metrology—these applications include stereo matching, learned stereo machines, phase unwrapping, calibration and point cloud analysis.

6.1 Introduction

Machine learning has been an active field of research since the 1960s (Samuel 1959, Nilsson 1965). However, for the majority of that time there have been few viable applications of machine learning techniques. With the emergence of high-powered central processing units, the immense parallel computing power of the Compute Unified Device Architecture (CUDA) accelerated graphics cards and more access to fully labelled training datasets through the advent of big data (Chang *et al* 2015, Garcia-Garcia *et al* 2017, Budianto *et al* 2019), many of the approaches developed through research can now be implemented commercially. Particular effort is being channelled into the areas of computer vision and image analysis (Tabernik *et al* 2020, Rehman *et al* 2019). While machine vision tasks have different requirements and often much less strict constraints when compared to measurement applications, there are many similarities to traditional form and coordinate metrology approaches—stereo vision analysis and scene segmentation in particular are active areas of research due to the advent of self-driving vehicles (Garcia-Garcia *et al* 2017). Machine learning techniques have already found success in different areas of metrology (Timoney *et al* 2020, Kong *et al* 2020, 2019, Barnes and Henn 2020), particularly for the measurement of very small parts such as semiconductors (Meng *et al* 2020, Henn *et al* 2019). Research groups and companies are investigating how these approaches can be adapted for use

doi:10.1088/978-0-7503-2524-0ch6

within optical metrology (Eastwood *et al* 2020). This chapter gives a brief introduction to machine learning techniques before providing a summary of current and developing applications to optical form and coordinate measurement, as well as techniques developed in other fields which could be applied to optical metrology.

6.2 Overview of machine learning and machine learning methods

While this chapter is not intended to be an instructional guide to machine learning (ML)—nor is it aimed at ML experts—it will be helpful to have a background understanding of the various ways ML can be implemented. This section will include a brief overview of the basic principles of ML before discussing artificial neural networks (ANNs), convolutional neural networks (CNNs) (Werbos 2018, Fukushima *et al* 1983) and support vector machines (SVMs) (Cortes and Vapnik 1995), which are the most common approaches to applying ML (Bishop 2006).

In general, ML can be thought of as a system which is not specifically programmed to solve a problem; it is instead told what problem to solve, given a set of training data, and then learns how best to solve the given problem on its own. More formally (Sun *et al* 2013, Valstar 2018), for an input space $\mathcal{X}$ and an output space $\mathcal{Y}$ the goal of ML is to learn a mapping function $h\colon \mathcal{X} \rightarrow \mathcal{Y}$, referred to as a hypothesis, from a dataset of labelled examples. The training dataset has entries $[\mathbf{x}, \mathbf{y}]$ where $\mathbf{x} \in \mathcal{X}$ are referred to as features and $f(\mathbf{x}) = \mathbf{y} \in \mathcal{Y}$ are referred to as labels. Training an ML model is the process of refining the hypothesis $h(\mathbf{x}) = \hat{\mathbf{y}}$, such that $\hat{\mathbf{y}}$ best approximates $\mathbf{y}$ and, thus, $h(\mathbf{x})$ approximates the unknown function $f(\mathbf{x})$. Once a hypothesis has been developed, this hypothesis can be used to make predictions on unseen data ($\mathbf{x}' \in \mathcal{X}$) that were not present in the initial dataset ($h(\mathbf{x}') = \hat{\mathbf{y}}'$). The accuracy of predictions on unseen data is referred to as the ability of the hypothesis to generalise, and the error between the set of all predictions and the corresponding ground-truth values ($\hat{\mathbf{y}}' - \mathbf{y}'$) is called the generalisation error. For large or continuous input spaces it is practically impossible to compute the full generalisation error as ML models can only be evaluated over a set of finite input samples, instead the generalisation error is estimated over a finite test dataset with elements ($\mathbf{x}'$, $\mathbf{y}'$). It is the aim of training ML models to minimise the generalisation error. This process is called supervised learning and contrasts with unsupervised learning where the initial dataset contains only features and no labels. Unsupervised learning is often referred to as pattern recognition, which is a separate but related approach to ML (Theodoridis *et al* 2010).

There are two main tasks completed by ML algorithms: categorisation and regression (Bishop 2006). Categorisation involves assigning each input datum to a discrete class, whereas regression produces a continuous value (or set of continuous values) for a given input. An example of classification is the detection of a part within an image; this could be a binary classification (*yes* there is an object, *no* there is not an object) or a multi-class classification (which object, of a set of possible objects, is present in the image). An example of a regression task is the pose estimation of an object within an image, where the ML model is regressing continuous coordinate values.

ANNs attempt to approximate the unknown function $f(\mathbf{x})$ by creating a highly dense, interconnected system of non-linear functions. Each individual branch (or neuron) in the system is simple, and through the large inter-connection of these simple components complex functions can be approximated. A basic ANN will look something like figure 6.1, albeit with a much larger number of nodes per layer.

The input nodes take the set of features, either from the training dataset ($\mathbf{x}$) or some unseen data ($\mathbf{x}'$). In an ANN, the features take the form of a tensor (n-dimensional vector) of numerical values. These data are passed to the hidden layers via a fully connected set of weighted connections (excluding the bias node in each hidden layer, shown in figure 6.1 as $h_{n,b}$). Each node in the hidden layers passes the weighted sum of these inputs through a non-linear activation function which determines the activation of that node. Each node in the next hidden layer takes the activations of the previous layer as input to the same process, producing complexity through linear combinations of non-linear activations. This is repeated through all the hidden layers in the given model until the final output nodes' activations are determined by the weighted sum of the activations in the final hidden layer. Figure 6.2 shows a detailed view of one of the neurons from the network shown in figure 6.1.

In order for the model to approximate complex functions, the activation function $a(z)$ must be non-linear. As a composition of linear functions can be described by a single linear function (i.e. $f_1(z) \circ f_2(z) = g(z)$ where g is a linear function for all f_1, f_2 which are linear), if $a(z)$ were linear this would allow the hidden layers in an ANN to be collapsed into a single hidden layer with a linear activation function, removing the ability of the network to capture complex functions. There are many possible candidate functions for $a(z)$; the most common are the rectified linear unit (ReLU) and the sigmoid (or logistic) function (Ramachandran *et al* 2017). The ReLU is described by

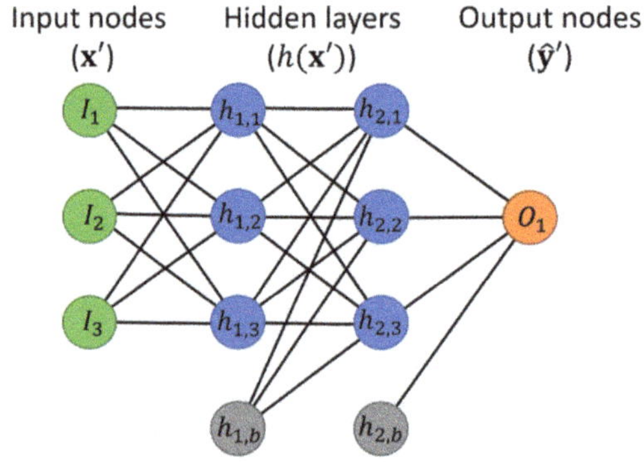

Figure 6.1. Basic ANN architecture. Input nodes shown in green, hidden nodes shown in blue (with bias nodes shown in grey) and output node shown in orange.

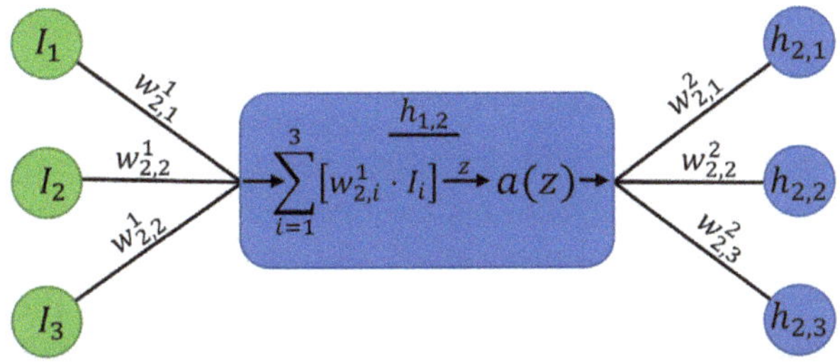

Figure 6.2. Detail of an ANN neuron (specifically $h_{1,2}$ from figure 6.1). Here $w^l_{j,k}$ is the weight of a connection, l, j and k are indices representing the layer, input node and output node, respectively, and $a(z)$ is a non-linear activation function which must be differentiable for all z.

$$a(z) = \begin{cases} z, & z > 0 \\ 0, & z \leqslant 0 \end{cases}. \tag{6.1}$$

In practice, the ReLU function is often approximated by other functions to allow it to be differentiable at $z = 0$. The logistic function is defined by

$$a(z) = \frac{e^z}{e^z + 1}. \tag{6.2}$$

By adjustment of the weights that connect together the nodes of an ANN, learning occurs. The adjustment of connection weights is done through a process of gradient descent and back propagation (Rumelhart *et al* 1986). Once the ANN has been designed (number of hidden layers, nodes per layer, activation function all selected), then the weights of each connection are initiated randomly. The first feature list from the training dataset is input to the network, fed forward through the layers, and then an output is generated. This output is then compared to the given label for that feature set through a loss function. The most common loss function for regression is the mean-squared error (MSE), while for classification, binary cross-entropy is commonly used and there are many other options (Janocha and Czarnecki 2017). The gradient of the selected loss function is calculated with respect to the node weights, which can then be adjusted in the direction which is expected to reduce the loss. This process is repeated until convergence is achieved. The weights are updated according to

$$\Delta w^l_{j,k} = \eta \cdot \frac{\partial E}{\partial a_{l,j}} \cdot \frac{\partial a_{l,j}}{\partial w^l_{j,k}}, \tag{6.3}$$

where w is the weight of a connection, l, j, k are indices representing the layer, input node and output node, respectively, E is the loss function, a is the activation function and η is a dimensionless coefficient called the learning rate, typically set around $\eta = 0.001$. The learning rate is used to control the step size taken during back propagation; many optimisation schemes (including the popular Adam optimiser (Kingma and Ba 2015)) use a variable learning rate which shrinks as training continues—forcing the optimisation to take smaller steps as convergence to the true value is approached.

CNNs, as ANNs, typically include a fully connected part of their network, however, they differ from ANNs by including an initial set of convolutional layers which act as input to the fully connected section (Krizhevsky *et al* 2012). The convolutional layers perform a process of sliding window kernel convolution, a process whereby a grid of values (or kernel) is convolved over the input to produce an output feature map. Depending on the size of the kernel and the step size taken across the pixel grid during convolution (called the stride), kernel convolution can also be used for spatial down-sampling of the input. It is common to use multiple layers of kernel convolution to produce more specific feature maps. This process is illustrated in figure 6.3.

As with the weights in the fully connected layers, the values within the kernel are learned through back propagation. A typical CNN will have a series of convolutional layers; each layer has a number of different kernel filters operating in parallel. The outputs of these filters are then fed into a fully connected section before producing the final prediction. Convolutional layers are often packaged together with batch-normalisation layers, which normalise the incoming data from the previous hidden layer, called pooling layers. Max-pooling takes a sliding window of inputs and outputs the maximum value, average-pooling does the same but outputs the mean value (pooling is often used for spatial down-sampling) and activation layers apply the previously discussed activation functions, such as ReLU, to the convolved data.

There are some variations on the simple CNN model. Fully convolutional neural networks (FCNNs) forego the fully connected layers and consist entirely of convolutional layers (Zhang *et al* 2019a). In recurrent neural networks (RNNs), each node takes as input not only the weighted sum of the previous layers, but also its own prior activation (Hochreiter and Schmidhuber 1997). In this way, RNNs are

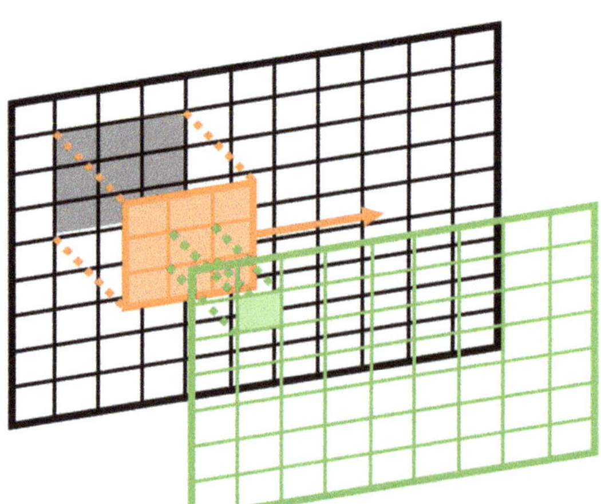

Figure 6.3. Kernel convolution. Input N-dimensional (shown as two-dimensional here) tensor shown in black, convolutional kernel shown in orange and output feature map tensor shown in green.

useful in dynamic systems where some form of memory is useful, or for applications such as language translation, where the previous word in a sentence may affect the meaning of future words. A residual neural network (ResNet) adds skip connections between layers so that each layer takes as input not only the activation of the previous layer, but also the layer before (He *et al* 2016). This mitigates against the 'vanishing gradient' problem in deep networks (Hochreiter 1998), whereby through repeated multiplication, the error gradient becomes so small that early layers are only updated by tiny amounts. The skip connections allow an alternative route for the gradient to propagate backwards through the network allowing deep networks to efficiently train their early layers.

While neural networks are popular, there are competing and complimentary approaches to ML that also see widespread application. SVMs are perhaps the most common alternative approaches. Simply, SVMs map training data to a higher-dimensional space where the categories are separated by a clear linear gap. Predictions can then be made on new data by mapping the unseen examples to the same space and determining which side of the gap they are mapped to. Neural networks are parametric approaches where, once the model is trained, the training dataset may be discarded, and predictions are made solely on the basis of the learned weights; in contrast, SVMs are memory-based. In this context, memory-based implies that the predictions are made on the basis of the data within the training set. SVMs are a type of kernel method which takes a set of data in an input space that are not linearly separable and maps them into a higher-dimensional space. These kernels are very different to the kernels used in sliding window convolution and are simply mapping functions of the input space into a new higher-dimensional space given by

$$k(\mathbf{x}, \mathbf{x}') = \phi(\mathbf{x})^T \cdot \phi(\mathbf{x}'), \tag{6.4}$$

where k is the kernel and ϕ is a feature map. SVMs employ the 'kernel trick' (Alpaydin 2020) which allows the kernel to be computed as a simple function in the input space without explicit knowledge of ϕ. For example, $k(\mathbf{x}, \mathbf{x}') = \mathbf{x} \cdot \mathbf{x}' + | \mathbf{x} |^2 | \mathbf{x}' |^2$ is a function that would be easily computable on a large set of input data and is equivalent to applying the feature map $\phi = (x, y, x^2 + y^2)$ as shown in equation (6.4). While this is only a three-dimensional example it is clear that the transformation can be completed without ever calculating an explicit representation of ϕ. This allows efficient computation of transforms into extremely high, sometimes infinite, dimensional spaces. Gaussian kernels are one of the most common kernel types and are an example of kernels that map the input space to an infinitely dimensional space (this can be seen through Taylor expansion of the Gaussian function, which leads to an infinite sum of inner products). When the data have been mapped to a space in which they are separable, a max-margin criterion is used to calculate the optimum hyper-plane to define the class boundary (Taskar *et al* 2004). Figure 6.4 shows two examples of how the same data can be separated, while figure 6.4(a) effectively separates the data and figure 6.4(b) uses the max-margin criteria to ensure the closest data point

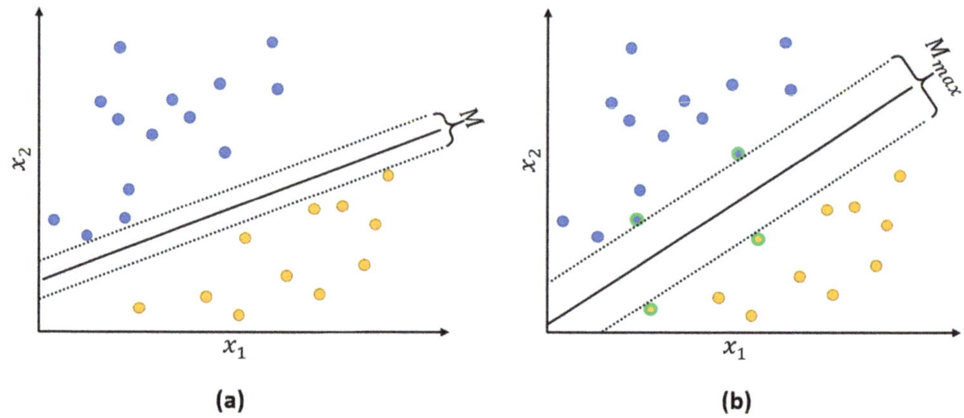

Figure 6.4. The same data separated by different hyper-planes (shown as a solid line). (a) The margin M (the distance from the plane to the nearest data) is not maximised. (b) The margin $M_{\max}$ is maximised and the data lying on the margin are defined as support vectors (highlighted in green).

to the hyper-plane from either class is a maximum distance ($M_{\max}$). The max--margin criteria decrease the probability of misclassification when the SVM is applied to unseen data.

The highlighted points (shown in figure 6.4(b) circled in green) are designated support vectors and are used to fully define the hyper-plane, therefore, when the SVM is deployed all other training data can be discarded and the support vectors alone are used to determine which class the new data belong to.

Many details were only summarised in this brief introduction, but this should provide a sufficient basis to understand the approaches discussed in the following sections, which cover the current state-of-the-art applications of ML to optical form and coordinate metrology.

6.3 Machine learning for stereo matching

A key stage in photogrammetric reconstruction, as discussed in chapter 4, is feature detection and feature matching between pairs of images, referred to here as 'stereo matching'. While the approaches discussed in chapter 4 can be highly effective, there has been recent effort to use ML to augment the stereo matching process (Li *et al* 2019, Kong and Tao 2006, Yin *et al* 2019c, Žbontar and Lecun 2016, Hu and Mordohai 2012, Motten *et al* 2011, Feng *et al* 2018, Haeusler *et al* 2013, Seki and Pollefeys 2016, Žbontar and Le Cun 2015) or to replace the traditional approaches entirely (Kar *et al* 2017, Becker and Hinton 1975, Li and Huttenlocher 2008, Fanello *et al* 2016, Kendall *et al* 2017, Xiang *et al* 2019). A common approach is to use ML to inform the computation of the stereo matching cost between detected features or local image patches. The multi-view stereo matching problem can be reformulated as a multi-class classification problem, where each class represents all possible views of a given feature (Lepetit *et al* 2004); this is shown to reduce the rate of matching errors. Competing approaches instead formulate the problem as a binary classification problem, where pairs of features are fed as inputs and the outputs represent the predicted probability that two features are matched. Early research in this area (Seki and Pollefeys 2016) takes as input two image patches which have been suggested as stereo matches by traditional methods. The two image patches are fed into a single CNN which has two output nodes predicting the probability that the suggested match is a correct match. An example of this type of CNN is shown in figure 6.5.

The CNN prediction is then directly used to adjust the semi-global stereo cost during reconstruction; this approach has been shown to appreciably improve depth estimation. More recently, it has become common in stereo vision applications to utilise Siamese networks (Luo *et al* 2016, Feng *et al* 2018, Žbontar and Le Cun 2015) that feed the paired input images into two separate copies of the same architecture, which are then compared to produce an output—this is illustrated in figure 6.6.

A Siamese network approach was implemented by Feng *et al* (2018), who used the predicted matching probability alone as the stereo matching cost. They showed that the CNN matching approach could outperform traditional approaches in some settings but did not produce accurate results for low-level texture and occluded surfaces. Therefore, they augment the purely ML approach with traditional algorithms, such as semi-global matching, interpolation, sub-pixel enhancement, median filtering and bilateral filtering.

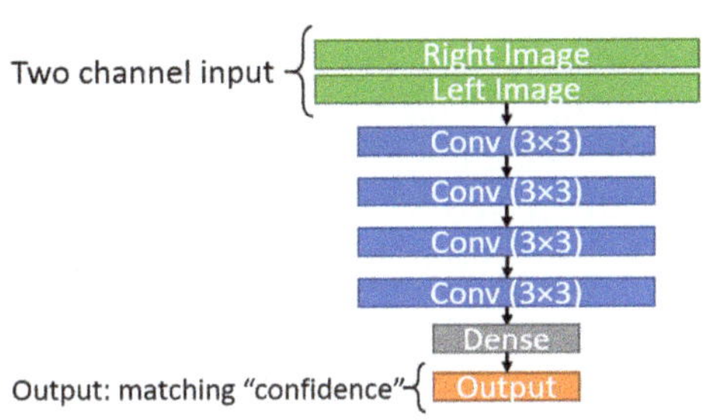

Figure 6.5. CNN to predict the 'confidence' that two stereo image patches are correctly matched. The convolutional layers (Conv) consist of kernel convolution with the shown kernel size, batch normalisation and ReLU activation.

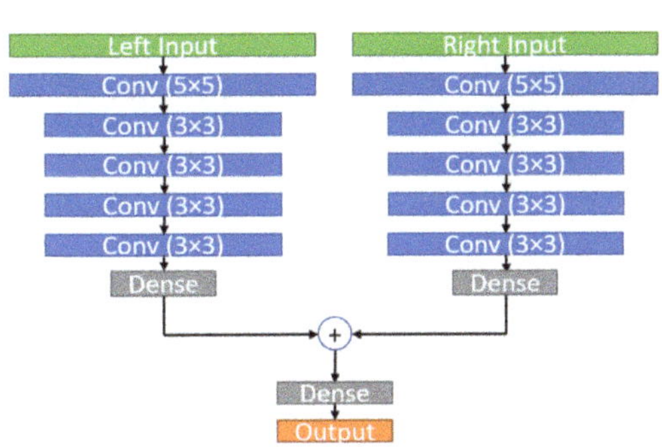

Figure 6.6. Example of a Siamese CNN for learned stereo matching. In contrast to figure 6.5, here the stereo pair are input to parallel copies of the same network with shared weights.

6.3.1 Learned stereo machines

Rather than augmenting traditional methods, some ML approaches seek to replace traditional algorithms entirely using end-to-end learning in so called learned stereo machines (LSMs). Learning stereography in this way is not a new idea—an early work (Becker and Hinton 1975) described how a neural network could be trained to discover depth from a pair of random dot stereograms of a curved surface—at the time this approach was limited to very simple images and surfaces, but with modern computing power can be expanded to more general applications. Work more recently (Kendall *et al* 2017) has begun to extend LSMs to create high-resolution depth maps from rectified image pairs, generating sub-pixel accuracy without the need for post-processing. Kar *et al* (2017) extend this idea from stereo pairs to multi-view systems. They propose a multi-view LSM which uses an RNN based approach to give two outputs, a three-dimensional (3D) voxel occupancy grid and per-view depth maps created through reprojection of the voxel data. A set of images of an object are passed to a feature detection CNN, these features are then un-projected into individual 3D spaces, an RNN is used to match the features and fuse the individual spaces into a single 3D world-space and a 3D CNN is then applied to this world-space to produce the final 3D prediction. This network architecture is shown in figure 6.7.

Using the approach shown in figure 6.7, reconstructions can be produced to a higher quality with fewer images than compared to traditional methods, however, the current implementation has a coarse voxel grid that makes it more suited to machine vision tasks than surface measurement. A more recent approach (Li and Huttenlocher 2008) uses a structured SVM to produce disparity maps from a stereo image pair. They show that an SVM based approach may produce significantly

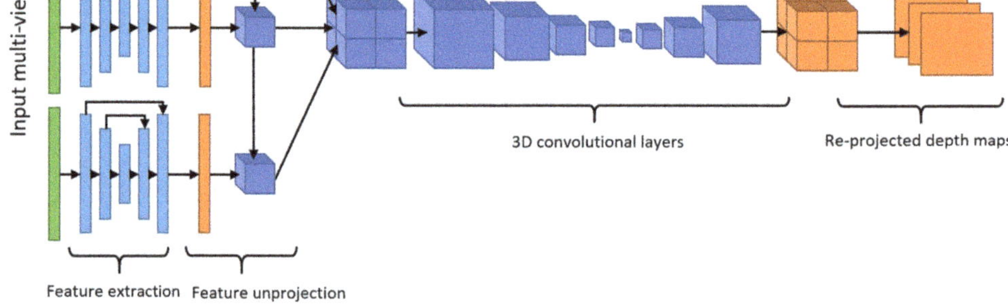

Figure 6.7. Overview of a learned stereo machine (LSM), based on Kar *et al* (2017), reprinted with permission from MDPI.

better performance than other learning-based methods when applied to the Middlebury-2005 dataset of stereo images (Scharstein and Pal 2007).

6.4 Machine learning for phase unwrapping

As discussed briefly in section 3.4.3, ML can be used to address the problem of phase unwrapping in fringe projection optical coordinate measuring systems (see chapter 5). Early work in this area used simple ANNs to perform phase unwrapping (Cuevas *et al* 2000), where they showed the potential for ML to recover fringe order at high speed with little knowledge of the camera characterisation or the details of the measurement device. However, this early work was prone to errors due to noise and variations in the surface characteristics between the training data and the measurement data. Recent work expands this ANN approach (Nguyen *et al* 2019) by encoding different fringe patterns in the RGB channels of a colour projector, which allows the camera to detect three individually phase-shifted patterns in a single-shot measurement. Corresponding pixels from the three wrapped phase maps are fed into the ANN, shown in figure 6.8, which can produce phase-unwrapped depth maps at 25.6 times a second, with a relative accuracy of 0.012%.

Rather than feed individual pixels through an ANN, there are many studies which use CNNs to address phase unwrapping (Feng *et al* 2019, 2020, Yin *et al* 2019b, 2019d, 2019a, Yang *et al* 2019, Budianto *et al* 2019, Van der Jeught and Dirckx 2019). Additionally, as coherence scanning interferometry (CSI, de Groot 2011, Su 2020) measurements contain the same phase unwrapping problem (albeit on a different scale), some techniques used in the context of CSI can also be applied to fringe projection (Zhang *et al* 2019b, Kando and Tomioka 2019, Zhong *et al* 2019).

Recent research uses an FCNN for phase unwrapping and some works (Zhang *et al* 2019b) treat the problem as a multi-class classification problem, where each class represents a fringe order, while alternative approaches (Van der Jeught and Dirckx 2019) directly regress depth values. The FCNN approach has the advantage that it can take images of different sizes without modification. In the multi-class

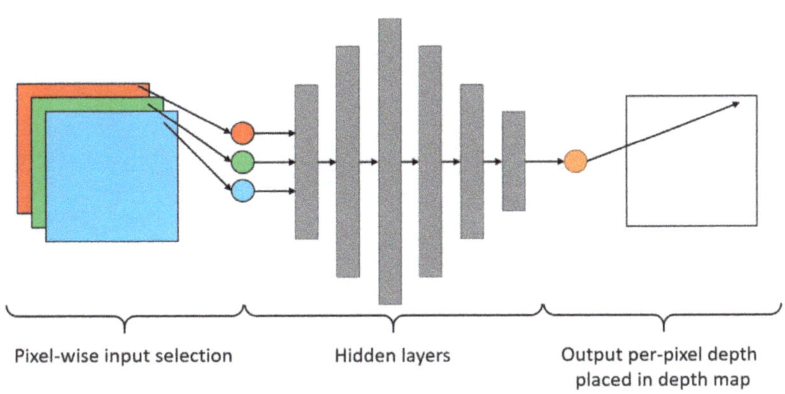

Figure 6.8. ANN for per-pixel phase unwrapping (Nguyen *et al* 2019), corresponding pixels from the three wrapped phase maps are passed through the network which produces a single output depth prediction.

approach (Zhang *et al* 2019b), an additional FCNN is implanted in order to pre-process noisy wrapped phase maps to remove or reduce the noise. Alternatively, an SVM with a radial basis function kernel can be used to classify each pixel in the wrapped phase map by its fringe order (Xiang *et al* 2020), also formulating the problem as a multi-class classification task. As this approach is per pixel, it avoids error propagation in spatial unwrapping and only a single projected fringe pattern is required, compared to the many projected fringe patterns used in traditional temporal phase unwrapping. It has been shown that using an SVM for phase unwrapping can produce depth maps of a comparable quality to traditional methods in a much shorter time. As with the CNN based methods, there is a competing SVR approach which formulates the problem as direct regression of depth, rather than a classification problem (Zhong *et al* 2019). Using SVR to directly predict depth values can produce measurement errors of under 1%, figure 6.9 shows a comparison of the SVR phase distribution, and the same measurement data unwrapped using a fast Fourier transform (FFT). It can be clearly seen in figure 6.9 that the SVR unwrapping outperforms the FFT unwrapping.

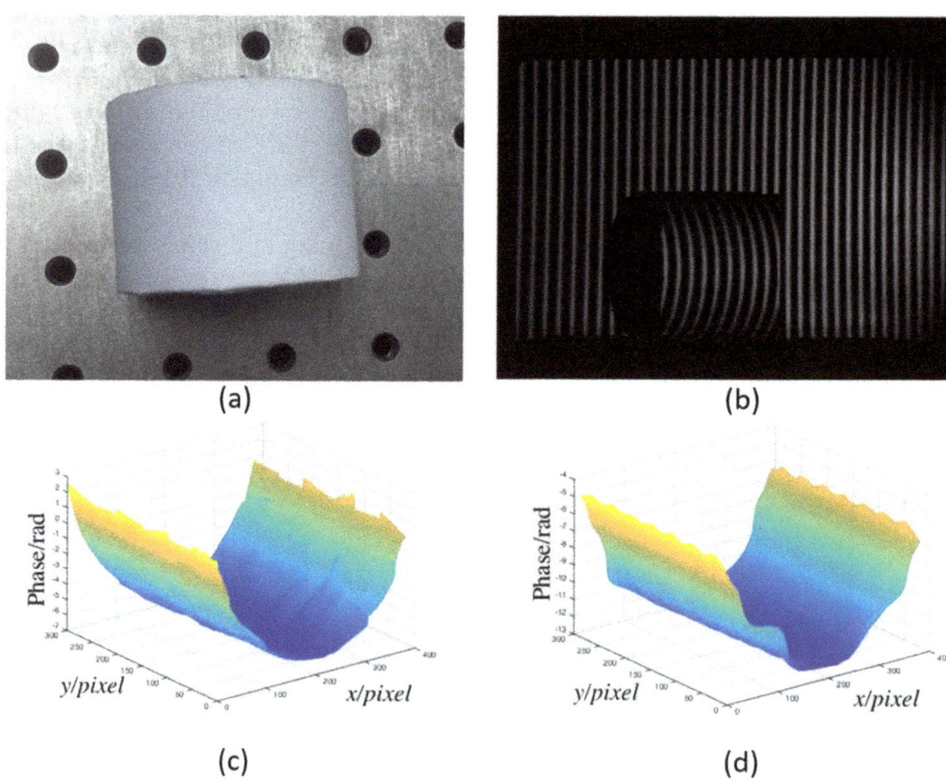

Figure 6.9. (a) The measured cylindrical object, (b) fringes projected onto the measured cylinder, (c) phase distribution calculated via SVR and (d) phase distribution calculated via FFT. Reproduced with permission from Zhong *et al* (2019).

6.5 Learning depth from a single image

Some fringe projection systems, and necessarily all photogrammetry systems, rely on multiple views to perform triangulation calculations for depth reconstruction. To simplify machine vision tasks, there has been recent research to infer depth information from a single image, using cues such as defocus and texture variation (Eigen *et al* 2014, Liu *et al* 2016). Some methods even produce point clouds (rather than depth maps or voxel grids) from a single image (Fan and Guibas 2017). Many of these approaches seem far too low resolution to be applicable to precision measurement applications, however, it has been shown that inferring depth from a set of single images can inform and improve reconstruction accuracy in multi-view systems (Trinh and McAllester 2009, Martins *et al* 2018). It is noted that many monocular cues, such as defocus and colour haze, apply even in regions of low texture where triangulation-based systems traditionally perform poorly (Saxena *et al* 2007). In this early work, it was demonstrated that incorporating both monocular and stereo cues produces higher accuracy depth maps than either approach could produce alone (Saxena *et al* 2007). More recent studies (Mansour *et al* 2019) discuss how monocular depth prediction can be applied to *in situ* monitoring—they demonstrate how monocular structure from motion depth estimates can improve the depth estimation range of a stereo camera pair. The structure from motion data based on the parallax and scene motion in the time step between images performs at a higher accuracy at greater distances from the sensor pair than the stereo triangulation data, which, in turn, shows greater accuracy at low distances from the sensors. Data fusion of both stereo and monocular cues can, therefore, produce the highest accuracy reconstructions over the greatest range of depths. By ascribing a confidence value to the depth prediction of each pixel, the optimum approach (stereo or monocular) can be selected on a per-pixel basis. When the stereo confidence value falls below 80%, the algorithm switches from using the stereo triangulation approach to the monocular method.

Efforts to improve monocular depth estimations can have direct implications for multi-view measurements. A recent example (Gur and Wolf 2019) focussed on purely using defocussing cues, by using a novel point spread function convolutional layer which applies location specific kernels to the input image. They use a supervised learning approach which outperforms similar unsupervised methods while achieving similar results to state-of-the-art supervised methods. The benefit of using an unsupervised approach is shown to be a better generalisation with data outside the training datasets than with the competing supervised methods. A state-of-the-art method (Garg *et al* 2019) exploits the dual-pixel design of modern cameras from which they claim 30% higher accuracy than any prior work on monocular depth estimation. To assist with auto-focus, many modern cameras utilise dual-pixel sensors, essentially splitting the green pixels into two via individual micro-lenses, similar to light field cameras. This hardware design can be exploited to create what is essentially a narrow-baseline stereo image pair. The extra information provided by the dual-pixel sensor allows more lightweight models to be implemented while maintaining high quality depth estimations. A pair of CNNs were used for training both via the Adam optimiser, one CNN based on the VGG architecture (Godard *et al* 2017) and the other based on U-Net (Ronneberger *et al* 2015), with the addition of residual layers (He *et al* 2016). In a comparison to the previous state-of-the-art approaches at the time their paper was published, they show a 30% accuracy improvement when using their approach on dual-pixel data. Machine learning for characterisation and calibration.

As the optical metrology industry moves towards a traceable calibration pipeline for measurement (see chapters 8 and 9), it becomes necessary to be able to quantify uncertainties related to any predictive ML models in the measurement pipeline (Vallejo *et al* 2019). Due to the complex nature of ML models, it is difficult to apply methodologies directly from the original *Guide to the Expression of Uncertainty in Measurement* (GUM, JCGM 2008a); rather a Monte-Carlo simulation approach must be taken (JCGM 2008b, Sediva and Havlikova 2013). Early work in this area showed how uncertainty can be propagated through simple linear regression models (Song *et al* 2013). Work by Cheung and Braun (2016) extended uncertainty analysis to more general model types and suggested that any analysis involving ML models should consider uncertainty contributions from:

- *Model output*: uncertainty relating to the difference between the model prediction and the ground-truth value.
- *Calibration data*: uncertainty in the data which make up the model training dataset.
- *Input measurement*: uncertainty in the input data to a model.
- *Output measurement*: output uncertainties outside the calibration of a dataset.

The study by Cheung and Braun demonstrates that increasing the size of the training dataset can reduce the uncertainty contribution from the calibration data but has no effect on the other contributions whose uncertainties are either from the derivation of the model, or inherent to the data themselves.

An example where ML has started to be applied is in the characterisation of a galvanometric laser scanner (Wissel *et al* 2015). In this study, the laser is controlled through a galvanometric positioned mirror, the mirror deflects the beam to scan over the surface being measured and the laser dot is then triangulated via a camera to a known position relative to the mirror. In characterising this system, Wissel *et al* showed that ML approaches can outperform model-based approaches and perform similarly to look-up table characterisation, providing coordinate root mean square errors (MSEs) as low as 0.029 mm and plane RMS errors of 0.433 mm (for a calibration plane size of 100 mm × 100 mm). They also show that using 'off-the-shelf' ANN architectures can lead to large generalisation errors and that reducing the problem using a specifically designed SVM can significantly reduce this error.

6.5.1 Characterisation of cameras and projectors

Although there has been some research to recreate depth from uncalibrated cameras (Lhuillier and Quan 2005), the characterisation of these parameters is key to the generation of accurate data for the application of measurement. Camera and projector characterisation are often referred to as 'calibration' in the literature, but from a metrological perspective it is really a process of characterisation as per the definitions given in chapter 1. Cameras and projectors can be described through a set of parameters that can be broken down into two categories: intrinsic and extrinsic. Intrinsic parameters are parameters internal to the camera/projector including focal length, principle point offset and skewness. Extrinsic parameters are external parameters such as position and orientation. Distortion parameters are also often included in the camera parametrisation. Commonly, low-order polynomials are used to model the radial lens distortion which can lead to a low-accuracy output. A new approach to characterising these distortion parameters is to use an SVM which has been shown to accurately capture distortion parameters with a height accuracy of 3 μm (Ergun *et al* 2010). It has been shown that CNNs can be more effective than conventional algorithms at detecting calibration checkerboards in photographic images (Raza *et al* 2019). Due to the requirement for multiple images at different angles, traditional calibration methods can be highly affected by low-accuracy components, such as rotation stages. There is potential to use deep CNNs to characterise a camera's intrinsic parameters from a single image (Lopez *et al* 2019), rather than the many images required for conventional approaches. Li and Liu (2019) used an MEMS micro-mirror device and a laser to stimulate single camera pixels and a deep CNN was then trained on this single pixel illumination data to calibrate the camera. Although this process may not save much time over the traditional methods, it can be automated to remove the dependence on the user. The single pixel illumination ML approach has high accuracy while requiring fewer computations than traditional calibration algorithms, producing an MSE of 0.0072 mm, which is an improvement of 60% over the traditional method.

Approaches for characterising the extrinsic parameters include the use of a CNN for real time estimation (Stavroulakis *et al* 2019). A CNN was trained to predict azimuthal and elevation angles from shadows cast by a single point source on a specific artefact. The CNN comprised five convolutional layers and a final fully connected layer. It was shown that there was an average $\pm 10°$ error in the azimuthal and elevation angles estimated which resulted in a ± 1 mm misalignment in the measured point cloud when compared to manually re-characterised extrinsic parameters.

6.6 Machine learning for point cloud analysis

As discussed in chapter 2, point cloud analysis is an important stage of any surface measurement pipeline as it is the stage at which information is extracted from the measurement data. Many machine learning-based approaches require the measured data to be regularised into a voxel grid because a voxel representation has a known size and known order. In contrast, point clouds do not have a regular ordered grid of points and the total number of points can vary from measurement to measurement. However, depending on the voxel grid resolution, transforming data into voxel form can either incur loss of detail or make the data size much larger. In order for an ML model to effectively operate directly on point clouds, it must fulfil the following criteria (Qi *et al* 2017a):

- For a point cloud of N 3D points, the model must be invariant under the $N!$ permutations of the order of the point cloud.
- The model must be able to capture local structure and interaction between neighbouring points.
- The model must be invariant under certain geometric transforms of the entire point cloud. For example, rotating the entire point cloud should not change the analysis result.

A simple approach to deal with inputs of an unknown size would be to use an RNN which iterates over each element in the input, however, it has been shown that the order of the RNN operating on the input directly effects the output, thus this approach does not satisfy the first requirement for operation on point clouds (Vinyals *et al* 2016). A better example using point clouds is given by PointNet (Qi *et al* 2017a), which is a neural network approach for point cloud segmentation discussed in the following section.

6.6.1 Point cloud segmentation

Segmentation of a point cloud is a widely researched area (see the reviews by Garcia-Garcia *et al* (2017), Grilli *et al* (2017) and Nguyen and Le (2013)) and, while most segmentation research is not explicitly conducted for use in measurement applications, there are many situations where it could apply, for example separating the measured object from the background, locating manufacturing defects and measuring multiple objects at the same time. Often the points are segmented into distinct semantic classes (such as background and object) in a process referred to as semantic segmentation. Semantic segmentation is distinct from instance segmentation, where semantic segmentation would simply label a point as either object or background, while instance segmentation also assigns a specific instance with each label, as shown in figure 6.10.

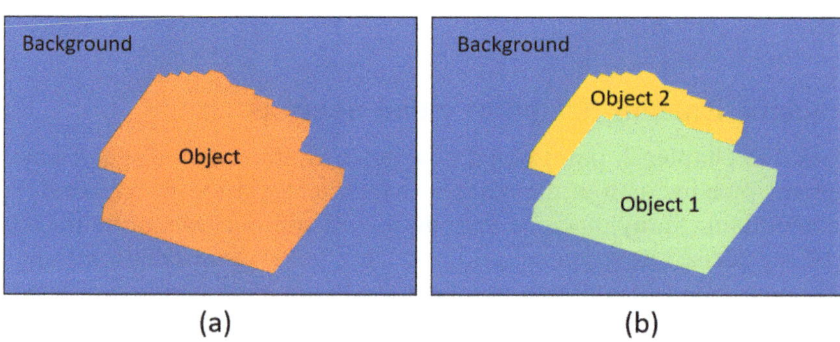

Figure 6.10. Semantic segmentation and instance segmentation of the same point cloud. (a) A point cloud semantically segmented into two classes, background and object. (b) A point cloud segmented into three classes, background and two distinct instances of the object.

An approach to segmentation is binary or multi-class classification on a per point basis. If the application can afford the sacrifices associated with using voxel representations, then 3D CNNs are well suited to this task (Huang and You 2016, Zhao *et al* 2019). For example, a point cloud can be transformed into a voxel representation by dividing the space into a grid and determining whether a voxel is occupied (contains at least one point) or is not occupied (Huang and You 2016). Figure 6.11 shows the network used to perform the multi-class classification problem. In their example, there were eight possible classes representing seven given categories and one class representing the case where the voxel belongs to no category (i.e. when it is not occupied).

As discussed, segmentation can also be applied directly to point clouds (Xu *et al* 2020, Qi *et al* 2017a, Garcia-Garcia *et al* 2017). PointNet (Qi *et al* 2017a) is essentially a series of ANNs connected together and notably has no convolutional layers. The architecture consists of two branches: the first classifies the entire cloud into one of k possible semantic classes. The global semantic classification is fed backwards into the second branch which classifies each point into one of m possible sub-classes. By concatenating the global classification result with the local feature maps, the local classification can be informed by both the local features and the global semantic class, satisfying the requirement that a model operating on point clouds can observe local structure. In order to satisfy the requirement for invariance over the order of the point cloud, a general function acting on the point cloud $f(\{x_1 \dots x_N\})$ is approximated by

$$f(\{x_1 \dots x_N\}) \approx g(h(x_1) \dots h(x_N)), \tag{6.5}$$

where $f\colon 2^{N\mathbb{R}} \to \mathbb{R}$, $h\colon \mathbb{R}^N \to \mathbb{R}^K$ and $g\colon \{\mathbb{R}_1^K \dots \mathbb{R}_N^K\} \to \mathbb{R}$ and is a symmetric function. An ANN used to approximate h and g is represented by a single variable function applied after pooling the values returned by h. By collecting multiple h, the properties of multiple functions f can be captured, and because g is symmetric the application of this approximation is invariant to input permutation. Figure 6.12 shows the architecture of PointNet where these features can be seen. To satisfy the final requirement that the model be invariant to geometric transformations on the entire point cloud, there are two small networks included in the model which both

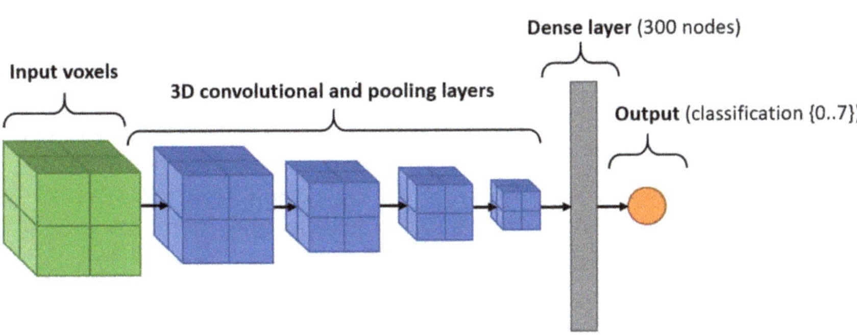

Figure 6.11. Multi-class voxel grid segmentation using a 3D CNN.

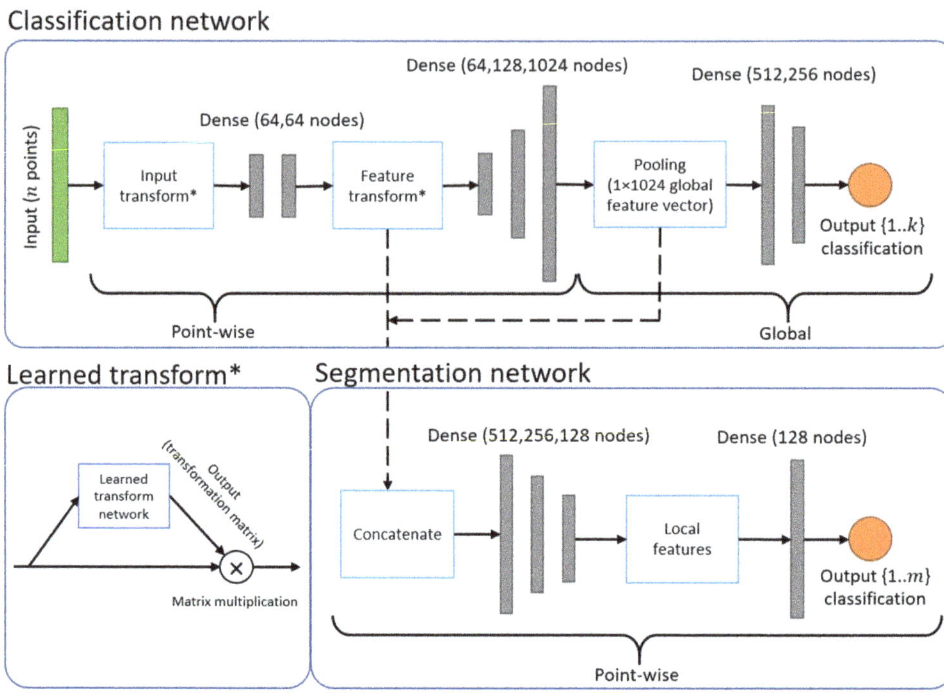

Figure 6.12. PointNet architecture (Qi *et al* 2017a). The classification network takes n input points, applies the learned input and feature transforms, pools the results and provides the global classification. The segmentation network takes the global result and local feature maps to perform per-point classification.

predict an affine transformation matrix. The first is applied to the entire input and the second is applied to the feature space (labelled input transform and feature transform, respectively, in figure 6.12).

PointNet has become an extremely popular approach due to its robust handling of pointsets, many of the upcoming examples use PointNet at some point in their model pipeline. In order to improve the performance of PointNet, specifically to capture local structure in the input point cloud, Qi *et al* (2017b) introduced a modification of PointNet called PointNet++. PointNet++ applies PointNet recursively on nested partitions of the input point cloud at a range of scales. This allows the learning of features to be robust to sampling density in the point cloud and has shown to be an improvement on the standard PointNet implementation on a set of benchmarks.

Segmentation is particularly useful for the detection and classification of surface defects (Tabernik *et al* 2020, Wen *et al* 2020a) and, if made fast enough, can be deployed for *in situ* monitoring (Tan *et al* 2020). This is an active area of research and also provides the motivation for much of the efficiency increases sought in both phase unwrapping (Liu *et al* 2020b) and stereo matching (Li *et al* 2018).

6.6.2 Point cloud registration

As covered in section 2.6.4, the registration of CAD data to a measured cloud or of two clouds together has many applications. A recent review by Zhang *et al* (2020) shows that numerous ML models have been proposed to replace many of the traditional approaches to point cloud registration. These models can be considered in two categories, models which aim to improve a single part of the registration pipeline (such as feature extraction), and models which seek to replace the entire pipeline with an end-to-end learning approach. An example of an approach which seeks to integrate into the traditional pipeline is the LORAX algorithm. The LORAX algorithm (Elbaz *et al* 2017) uses a deep ANN for dimensionality reduction, to attempt to simplify the registration problem. Essentially, the algorithm detects features in the point clouds, compares them for semantic similarity, matches features between the clouds (hence, coarsely aligning the two clouds) and finally refines the alignment with the iterative closest point (ICP) algorithm (see chapter 2). Comparing features can be computationally expensive if the dimensionality of the feature space is large, therefore, an ANN is used to create a compact representation of each feature and these representations are compared for similarity. The ANN is set up as an auto-encoder, which is shown in figure 6.13.

The encoder ANN takes the input and outputs a compact vector. The decoder takes this vector and attempts to recreate the input. By comparing the decoded input with the actual input, a model loss can be derived which determines how well the compact representation captures the input. This allows unsupervised training to be conducted, resulting in a network that can take large features and produce a vector which represents that feature in a much more compact manner, allowing for more efficient matching. Related research, called deep closest point (Wang and Solomon 2019), has also shown that using learned features to provide initial alignment between two point clouds can greatly facilitate accurate alignment via ICP, and can make ICP far more robust to failure when dealing with point clouds with poor initial alignment. As stated earlier, other approaches, such as DeepVCP (Lu *et al* 2019), attempt to fully replace approaches such as RANSAC (as discussed in section

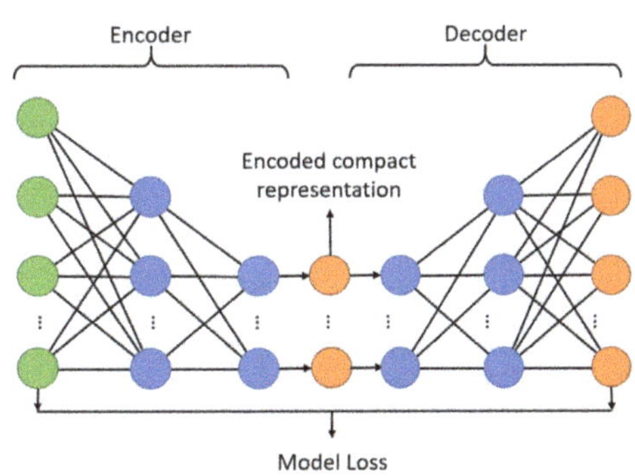

Figure 6.13. Auto-encoder architecture. Training is conducted such that the output can be decoded from the compact representation with minimum difference when compared to the input, therefore, ensuring the compact representation captures the input as fully as possible.

2.5.1.5) and ICP with an end-to-end learned model. An example of a fully ML based approach to registration for form measurement is research by Gojcic *et al* (2020), which focuses on combining multi-view point clouds into a completed single point cloud.

Figure 6.14 shows the approach to multi-view point cloud registration which is outlined here—for each input point cloud, a neural network extracts a set of features. These features are fed into a further set of network blocks that compute stochastic correspondences between each combination of pairs of point clouds. Using these correspondences, a further block (labelled Reg. init. in figure 6.14) computes initial transformation parameters and residuals, which are refined by the next block (Reg. iter.) from which the registration graph is built. The final two blocks are iterated four times to produce the complete registration. This approach is evaluated on a set of benchmark datasets and shown to outperform competing approaches on the same datasets by 25% in terms of mean rotational error. Furthermore, the approach is shown to be thirteen times faster than RANSAC (see chapter 2) when registering a set of 60 point clouds and copes well with unseen scenes and objects.

In the conclusion of their review, Zhang *et al* (2020) make the following points: ML models show clear dominance when used as a module within a traditional pipeline, such as for feature extraction, and end-to-end learned systems are growing in popularity and have the potential to become more effective than traditional methods. They also note that some failings of current ML models include the existence of a clear gap between performance on synthetic datasets and performance on real world data. A further area for improvement is that many systems use feature-extraction networks which were designed for other applications, such as PointNet and PointNet++, there is likely to be a performance gain by designing a bespoke approach for registration.

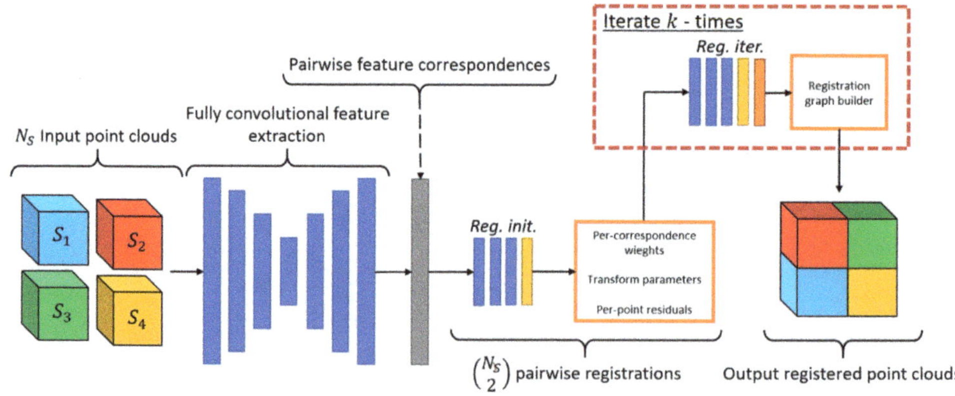

Figure 6.14. ML approach for multi-view point cloud registration (Gojcic *et al* 2020). Features are extracted from the input point clouds and then iteratively matched to find the best alignment.

6.6.3 Point cloud completion

Often a measurement will result in an incomplete point cloud due to occluded surfaces, either due to insufficient views or surface complexity. It can, therefore, be useful to attempt to predict the missing data, not for extending the measurement data, but for increasing the performance of registration algorithms such as ICP. It is common to use an auto-encoder type network and RNNs as part of the pipeline for point cloud completion (Liu *et al* 2020a, Wen *et al* 2020b, Van Gansbeke *et al* 2019, Giancola *et al* 2019). Figure 6.15 shows an example approach for point cloud completion (Liu *et al* 2020a).

The approach shown in figure 6.15 yielded successful results, and the point sampling approach used ensures that any known structures from the incomplete point cloud are preserved. It is not clear if this method will generalise well onto unseen objects of different semantic classes to those contained within the training dataset and the supervised learning approach also requires a large labelled training dataset. A similar approach uses PointNet, which was described in figure 6.12, as the layers in its encoder (Wen *et al* 2020b), this has similar trade-offs as using PointNet for point cloud registration as was described in the previous section.

Another approach to point cloud completion is to use a generative adversarial network (GAN) (Sarmad *et al* 2019, Chen *et al* 2019). A GAN can be thought of as two competing networks, referred to as a generator and a discriminator. The generator attempts to create new data (in this case, complete point clouds) from an input latent-space vector. The discriminator is trained on a combined dataset of real complete and generated point clouds, and is trained to predict whether a given input is real or generated. The GAN sequences these two networks and the discriminator output is used as the loss of the generator. Thus the generator is trained to produce point clouds that the discriminator cannot distinguish from real data. GANs can be used directly on input point clouds, do not require labelled training data and allow more general application of the network to make predictions on unseen data from unknown semantic classes (Chen *et al* 2019). An extra reinforcement learning step can be inserted into the GAN pipeline (called RL-GAN-Net (Sarmad *et al* 2019)) which learns how controlling the input vector affects the output point cloud. This

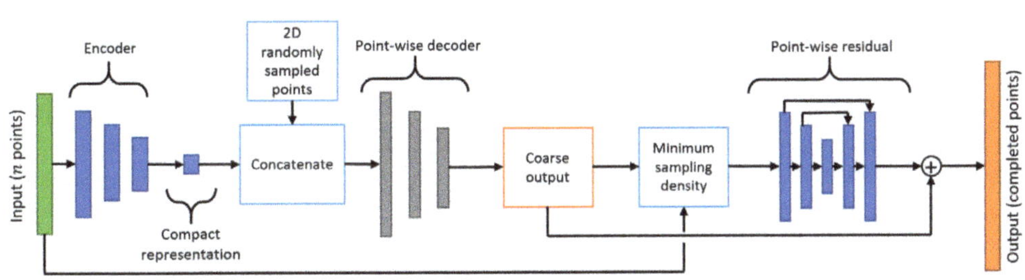

Figure 6.15. Point cloud completion network Liu *et al* (2020a). This approach to point cloud completion uses an encoder to create a compact representation of the input, this is then combined with a random 2D set of points to predict a mapping of those points onto unknown 3D surfaces, a set of these predicted surface are combined to create a coarse completion estimation. This estimation is then further densified through fed-forward information from the input and ResNet layers.

can, therefore, be trained to provide an input which is likely to produce an output cloud which is most similar to the missing data. RL-GAN-Net can generate completed point clouds in approximately 1 ms, allowing it to be deployed in real time situations and produce completed point clouds from input data with up to 70% missing points.

6.7 Conclusions

While the application of machine learning remains a developing field, there are many areas where machine learning models can be usefully applied to optical form and coordinate metrology. Stereo matching and phase unwrapping have both been shown to benefit from the integration of machine learning into existing algorithms, and from a complete end-to-end learning approach that, in many cases, outperforms the traditional approaches. Calibration and characterisation remain open problems —calibration in particular requires much more research into the uncertainty contributions generated by machine learning models in a measurement pipeline. Characterisation of cameras has the potential to be automated through machine learning processes, but further investigation is still required for this technology to outperform traditional characterisation techniques. Analysis of point clouds is a further area where machine learning certainly has a role to play. Segmentation tasks are an example of where machine learning can provide results which would be unachievable otherwise. The registration of point clouds to computer aided design models for inspection is also an active area of research for machine learning, often used in a pipeline with iterative techniques such as the iterative closest point algorithm. While much of the literature requires point clouds to be converted into voxel representations, which is not ideal for measurement applications, and many machine learning models only operate on very low voxel resolutions, a growing body of work operates directly on high-fidelity point clouds. Many of the studies focusing on voxel representations, therefore, could be generalised through further research to operate on point clouds, meaning much of the work which currently refers to voxel grids may become relevant for measurement applications.

References

Alpaydin E 2020 *Introduction to Machine Learning* (Cambridge, MA: MIT Press)

Barnes B M and Henn M-A 2020 Contrasting conventional and machine learning approaches to optical critical dimension measurements *Proc. SPIE* **11325** 113251E

Becker S and Hinton G E 1975 Self-organizing neural network that discovers surfaces in random-dot stereograms *Nature* **255** 242–3

Bishop C M 2006 *Pattern Recognition and Machine Learning* (Berlin: Springer)

Budianto B, Law W and Lun D P K 2019 Deep learning based period order detection in structured light three-dimensional scanning *Proc. IEEE ISCAS* pp 0–4

Chang A X *et al* 2015 ShapeNet: an information-rich 3D model repository arXiv: 1512.03012

Chen X, Chen B and Mitra N J 2019 Unpaired point cloud completion on real scans using adversarial training arXiv: 1904.00069

Cheung H and Braun J E 2016 A general method for calculating the uncertainty of virtual sensors for packaged air conditioners *Int. J. Refrig* **63** 225–36

Cortes C and Vapnik V 1995 Support-vector networks *Mach. Learn.* **20** 273–97

Cuevas F J, Servin M, Stavroudis O N and Rodriguez-Vera R 2000 Multi-layer neural network applied to phase and depth recovery from fringe patterns *Opt. Commun.* **181** 239–59

de Groot P 2011 Coherence scanning interferometry *Optical Measurement of Surface Topography* ed R K Leach (Berlin: Springer) pp 187–208

Eastwood J, Zhang H, Isa M, Sims-Waterhouse D, Leach R K and Piano S 2020 Smart photogrammetry for three-dimensional shape measurement *Proc. SPIE* **11352** 113520A

Eigen D, Puhrsch C and Fergus R 2014 Depth map prediction from a single image using a multi-scale deep network *Proc. NIPS (Montréal, Canada) vol 3 pp 2366–74*

Elbaz G, Avraham T and Fischer A 2017 3D point cloud registration for localization using a deep neural network auto-encoder *Proc. IEEE CVPR* pp 2472–81

Ergun B, Kavzoglu T, Colkesen I and Sahin C 2010 Data filtering with support vector machines in geometric camera calibration *Opt. Express* **18** 1927

Fan H and Guibas L 2017 DeepPointSet: a point set generation network for 3D object reconstruction from a single image *Proc. IEEE CVPR* pp 1–4

Fanello S R, Rhemann C, Tankovich V, Kowdle A, Escolano S O, Kim D and Izadi S 2016 HyperDepth: learning depth from structured light without matching *Proc. IEEE CVPR* pp 5441–50

Feng S, Zuo C, Chen Q and Gu G 2019 Fringe analysis based on convolutional neural networks *Proc. SPIE* **10991** 109910C

Feng S, Zuo C, Yin W and Chen Q 2020 Application of deep learning technology to fringe projection 3D imaging *Infrared Laser Eng.* **49** 1–17

Feng Y, Liang Z and Liu H 2018 Efficient deep learning for stereo matching with larger image patches *Proc. IEEE CISP-BMEI (Shanghai, China)* pp 1–5

Fukushima K, Miyake S and Ito T 1983 Neocognitron: a neural network model for a mechanism of visual pattern recognition *Competition and Cooperation in Neural Nets* ed S-I Amari and M A Arbib (Berlin: Springer)

Garcia-Garcia A, Orts-Escolano S, Oprea S, Villena-Martinez V and Garcia-Rodriguez J 2017 A review on deep learning techniques applied to semantic segmentation *Appl. Soft Comput.* **70** 41–65

Garg R, Wadhwa N, Ansari S and Barron J T 2019 Learning single camera depth estimation using dual-pixels *Proc. IEEE ICVF (Seoul, South Korea)* pp 7628–37

Giancola S, Zarzar J and Ghanem B 2019 Leveraging shape completion for 3D Siamese tracking *Proc. IEEE CVPR* pp 1359–68

Godard C, Mac Aodha O and Brostow G J 2017 Unsupervised monocular depth estimation with left-right consistency *Proc. IEEE CVPR* pp 270–9

Gojcic Z, Zhou C, Wegner J D, Guibas L J and Birdal T 2020 Learning multiview 3D point cloud registration *Proc. IEEE CVPR* pp 1759–69

Grilli E, Menna F and Remondino F 2017 A review of point clouds segmentation and classification algorithms *Int. Arch. Photogramm. Remote Sens. Spat. Inf. Sci.* **42** 339–44

Gur S and Wolf L 2019 Single image depth estimation trained via depth from defocus cues *Proc. IEEE CVPR* pp 7683–92

Haeusler R, Nair R and Kondermann D 2013 Ensemble learning for confidence measures in stereo vision *Proc. IEEE CVPR* pp 305–12

He K, Zhang X, Ren S and Sun J 2016 Deep residual learning for image recognition *Proc. IEEE CVPR* pp 770–8

Henn M-A, Zhou H, Silver R M and Barnes B M 2019 Applications of machine learning at the limits of form-dependent scattering for defect metrology *Proc. SPIE* **10959** 109590Z

Hochreiter S 1998 The vanishing gradient problem during learning recurrent neural nets and problem solutions *Int. J. Uncertainty Fuzziness Knowledge-Based Syst.* **6** 107–16

Hochreiter S and Schmidhuber J 1997 Long short-term memory *Neural Comput.* **9** 1735–80

Hu X and Mordohai P 2012 A quantitative evaluation of confidence measures for stereo vision *IEEE Trans. Pattern Anal. Mach. Intell.* **34** 2121–33

Huang J and You S 2016 Point cloud labeling using 3D convolutional neural network *Proc. IEEE ICPR (Cancun, Mexico)* pp 2670–5

Janocha K and Czarnecki W 2017 On loss functions for deep neural networks in classification *Schedae Inform.* **25** 49–59

JCGM 100 2008a *GUM 1995 with Minor Corrections, Evaluation of Measurement Data—Guide to the expression of Uncertainty in Measurement* (Sèvres: BIPM)

JCGM 101 2008b *Evaluation of Measurement Data—Supplement 1 to the Guide to the Expression of Uncertainty in Measurement—Propagation of Distributions Using a Monte Carlo Method* (Sèvres: BIPM)

Kando D and Tomioka S 2019 Phase extraction from interferogram using machine learning *Proc. EI2019 (San Francisco, CA, USA)* pp 1–5

Kar A, Häne C and Malik J 2017 Learning a multi-view stereo machine *Proc. NIPS (Long Beach, CA, USA)* pp 365–76

Kendall A, Martirosyan H, Dasgupta S, Henry P, Kennedy R, Bachrach A and Bry A 2017 End-to-end learning of geometry and context for deep stereo regression *Proc. IEEE ICCV (Venice, Italy)* pp 66–75

Kingma D P and Ba J 2015 Adam: a method for stochastic optimization *Proc. ICLR (San Diego, CA, USA)*

Kong D, Motoyama K, Huang H, Mendoza B, Breton M, Muthinti G R, Shobha H, Jiang L, Li J and Demarest J J 2019 Machine learning and hybrid metrology using scatterometry and LE-XRF to detect voids in copper lines *Proc. SPIE* **10959** 109590A

Kong D, Schmidt D, Church J, Liu C, Breton M, Murray C, Miller E, Meli L, Sporre J and Felix N 2020 Measuring local CD uniformity in EUV vias with scatterometry and machine learning *Proc. SPIE* **11325** 113251I

Kong D and Tao H 2006 Stereo matching via learning multiple experts behaviors *Proc. BMVC (Edinburgh, UK)* pp 97–106

Krizhevsky A, Sutskever I and Hinton G E 2012 ImageNet classification with deep convolutional neural networks *Proc. NIPS (Lake Tahoe, NV)* 1097–105

Lepetit V, Pilet J and Fua P 2004 Point matching as a classification problem for fast and robust object pose estimation *Proc. IEEE CVPR* vol 2 2

Lhuillier M and Quan L 2005 A quasi-dense approach to surface reconstruction from uncalibrated images *IEEE Trans. Pattern Anal. Mach. Intell.* **27** 418–33

Li F, Li Q, Zhang T, Niu Y and Shi G 2019 Depth acquisition with the combination of structured light and deep learning stereo matching *Signal Process. Image Commun.* **75** 111–7

Li J and Liu Z 2019 Camera geometric calibration using dynamic single-pixel illumination with deep learning networks *IEEE Trans. Circuits Syst. Video Technol.* **30** 2550–8

Li Y and Huttenlocher D P 2008 Learning for stereo vision using the structured support vector machine *Proc. IEEE CVPR* pp 1–8

Li Z, Liu X, Wen S, He P, Zhong K, Wei Q, Shi Y and Liu S 2018 *In situ* 3D monitoring of geometric signatures in the powder-bed-fusion additive manufacturing process via vision sensing methods *Sensors* **18** 1180

Liu F, Shen C, Lin G and Reid I 2016 Learning depth from single monocular images using deep convolutional neural fields *IEEE Trans. Pattern Anal. Mach. Intell.* **38** 2024–39

Liu M, Sheng L, Yang S, Shao J and Hu S-M 2020a Morphing and sampling network for dense point cloud completion *Proc. AAAI-20 (New York, USA) vol. 34 pp 11596–603*

Liu Y, Blunt L, Zhang Z, Rahman H A, Gao F and Jiang X 2020b *In-situ* areal inspection of powder bed for electron beam fusion system based on fringe projection profilometry *Addit. Manuf.* **31** 100940

Lopez M, Mari R, Gargallo P, Kuang Y, Gonzalez-Jimenez J and Haro G 2019 Deep single image camera calibration with radial distortion *Proc. IEEE CVPR* pp 11817–25

Lu W, Wan G, Zhou Y, Fu X, Yuan P and Song S 2019 DeepVCP: an end-to-end deep neural network for point cloud registration *Proc. IEEE ICCV (Seoul, South Korea)* pp 12–21

Luo W, Schwing A G and Urtasun R 2016 Efficient deep learning for stereo matching *Proc. IEEE CVPR* pp 5695–703

Mansour M, Davidson P, Stepanov O and Piché R 2019 Relative importance of binocular disparity and motion parallax for depth estimation: a computer vision approach *Remote Sens.* **11** 1990

Martins D, Van Hecke K and De Croon G 2018 Fusion of stereo and still monocular depth estimates in a self-supervised learning context *Proc. IEEE ICRA (Brisbane, Australia)* 849–56

Meng M, Tu L, Mi J, Zhou H and Zou X 2020 Machine learning and hybrid metrology using HV-SEM and optical methods to monitor channel hole tilting in-line for 3D NAND wafer production *Proc. SPIE* **11325** 113250O

Motten A, Claesen L and Pan Y 2011 Binary confidence evaluation for a stereo vision based depth field processor SoC *Proc. ACPR (Beijing, China)* pp 456–60

Nguyen A and Le B 2013 3D point cloud segmentation: a survey *Proc. IEEE RAM (Manila, Philippines)* pp 225–30

Nguyen H, Dunne N, Li H, Wang Y and Wang Z 2019 Real-time 3D shape measurement using 3LCD projection and deep machine learning *Appl. Opt.* **58** 7100

Nilsson N J 1965 *Learning Machines* (New York: McGraw-Hill)

Qi C R, Su H, Mo K and Guibas L J 2017a PointNet: deep learning on point sets for 3D classification and segmentation *Proc. IEEE CVPR* pp 652–60

Qi C, Yi L, Su H and Guibas L 2017b PointNet++: deep hierarchical feature learning on point sets in a metric space *Proc. NIPS (Long Beach, CA, USA)* pp 5105–14

Ramachandran P, Zoph B and Le Q V 2017 Searching for activation functions arXiv: 1710.05941

Raza S N, ur Rehman H R, Lee S G and Choi G S 2019 Artificial intelligence based camera calibration *Proc. IWCMC (Tangier, Morocco)* pp 1564–9

Rehman T U, Mahmud M S, Chang Y K, Jin J and Shin J 2019 Current and future applications of statistical machine learning algorithms for agricultural machine vision systems *Comput. Electron. Agric.* **156** 585–605

Ronneberger O, Fischer P and Brox T 2015 U-Net: convolutional networks for biomedical image segmentation *Proc. MICCAI (Munich, Germany)* pp 234–41

Rumelhart D E, Hinton G E and Williams R J 1986 Learning representations by back-propagating errors *Nature* **323** 533–6

Samuel A L 1959 Some studies in machine learning using the game of checkers *IBM J. Res. Dev.* **3** 210–29

Sarmad M, Lee H J and Kim Y M 2019 RL-GAN-net: a reinforcement learning agent controlled GAN network for real-time point cloud shape completion *Proc. IEEE CVPR* pp 5891–900

Saxena A, Jamie S and Ng A Y 2007 Depth estimation using monocular and stereo cues *Proc. IJCAI (Hyderabad, India)* pp 2197–203

Scharstein D and Pal C 2007 Learning conditional random fields for stereo *Proc. IEEE CVPR* pp 1–8

Sediva S and Havlikova M 2013 Comparison of GUM and Monte Carlo method for evaluation measurement uncertainty of indirect measurements *Proc. ICCC (Rytro, Poland)* pp 325–9

Seki A and Pollefeys M 2016 Patch based confidence prediction for dense disparity map *Proc. BMVS (York, UK) vol. 23 pp 1–13*

Song L, Wang G and Brambley M R 2013 Uncertainty analysis for a virtual flow meter using an air-handling unit chilled water valve *HVAC&R Res.* **19** 335–45

Stavroulakis P, Chen S, Delorme C, Bointon P, Tzimiropoulos G and Leach R K 2019 Rapid tracking of extrinsic projector parameters in fringe projection using machine learning *Opt. Lasers Eng.* **114** 7–14

Su R 2020 Coherence scanning interferometry *Advances in Optical Surface Texture Metrology* ed R K Leach (Bristol: IOP Publishing)

Sun L, Ji S and Ye J 2013 *Multi-Label Dimensionality Reduction* (Boca Raton, FL: CRC Press)

Tabernik D, Šela S, Skvarč J and Skočaj D 2020 Segmentation-based deep-learning approach for surface-defect detection *J. Intell. Manuf.* **31** 759–76

Tan Z, Fang Q, Li H, Liu S, Zhu W and Yang D 2020 Neural network based image segmentation for spatter extraction during laser-based powder bed fusion processing *Opt. Laser Technol.* **130** 106347

Taskar B, Klein D, Colllins M, Koller D and Manning C D 2004 Max-margin parsing *Proc. EMNLP (Barcelona, Spain)* pp 1–8

Theodoridis S, Pikrakis A, Koutroumbas K and Cavouras D 2010 *Introduction to Pattern Recognition: A Matlab Approach* (New York: Academic)

Timoney P, Luthra R, Elia A, Liu H, Isbester P, Levy A, Shifrin M, Bringoltz B, Rabinovich E and Broitman A 2020 Advanced machine learning eco-system to address HVM optical metrology requirements *Proc. SPIE* **11325** 113251H

Trinh H and McAllester D 2009 Unsupervised learning of stereo vision with monocular cues *Proc. BMVC (London, UK) vol 4 pp 1–11*

Vallejo M, De La Espriella C, Gómez-Santamaría J, Ramírez-Barrera A F and Delgado-Trejos E 2019 Soft metrology based on machine learning: a review *Meas. Sci. Technol.* **31** 032001

Valstar M 2018 *Introduction lecture notes Machine Learning COMP3009* University of Nottingham, delivered 16 October 2018

Van der Jeught S and Dirckx J J J 2019 Deep neural networks for single shot structured light profilometry *Opt. Express* **27** 17091

Van Gansbeke W, Neven D, De Brabandere B and Van Gool L 2019 Sparse and noisy lidar completion with RGB guidance and uncertainty *Proc. IEEE MVA (Tokyo, Japan)* pp 1–6

Vinyals O, Bengio S and Kudlur M 2016 Order matters: sequence to sequence for sets *Proc. ICLR (San Juan, Puerto Rico)* pp 1–11

Wang Y and Solomon J 2019 Deep closest point: learning representations for point cloud registration *Proc. IEEE ICCV (Seoul, South Korea)* pp 3522–31

Wen V, Wong H, Ferguson M, Law K H, Lee Y-T T and Witherell P 2020a Automatic volumetric segmentation of additive manufacturing defects with 3D U-Net *Proc. Spring Symp. AI Manuf. (Palo-Alto, CA, USA)*

Wen X, Li T, Han Z and Liu Y-S 2020b Point cloud completion by skip-attention network with hierarchical folding *Proc. IEEE CVPR* pp 1939–48

Werbos P 2018 Beyond regression: new tools for prediction and analysis in the behavioural scieces *PhD Dissertation* Harvard University, Cambridge, MA

Wissel T, Wagner B, Stüber P, Schweikard A and Ernst F 2015 Data-driven learning for calibrating galvanometric laser scanners *IEEE Sens. J.* **15** 5709–17

Xiang S, Yang Y, Deng H, Wu J and Yu L 2019 Multi-anchor spatial phase unwrapping for fringe projection profilometry *Opt. Express* **27** 33488

Xiang S, Deng H, Wu J and Zhu C 2020 Absolute phase unwrapping with SVM for fringe-projection profilometry *IET Image Process.* **14** 2645–51

Xu M, Zhou Z and Qiao Y 2020 Geometry sharing network for 3D point cloud classification and segmentation *Proc. AAAI (New York, USA)* pp 12500–7

Yang C, Yin W, Xu H, Li J, Feng S, Tao T, Chen Q and Zuo C 2019 Single-shot 3D shape measurement with spatial frequency multiplexing using deep learning *Proc. SPIE* **11189** 111891P

Yin W, Chen Q, Feng S, Tao T, Huang L, Trusiak M, Asundi A and Zuo C 2019a Temporal phase unwrapping using deep learning *Sci. Rep.* **9** 1–12

Yin W, Zuo C, Feng S, Tao T and Chen Q 2019b Bi-frequency temporal phase unwrapping using deep learning *Proc. SPIE* **10991** 109910D

Yin W, Zuo C, Feng S, Tao T and Chen Q 2019c High-speed 3D shape measurement with the multi-view system using deep learning *Proc. SPIE.* **11189** 111890B

Yin W, Zuo C, Feng S, Tao T and Chen Q 2019d Temporal phase unwrapping using multi-scale deep neural networks *Proc. SPIE* **11189** 111890A

Žbontar J and Le Cun Y 2015 Computing the stereo matching cost with a convolutional neural network *Proc. IEEE CVPR* pp 1592–9

Žbontar J and Le Cun Y 2016 Stereo matching by training a convolutional neural network to compare image patches *J. Mach. Learn. Res.* **17** 1–32

Zhang A, Lipton Z C, Li M and Smola A J 2019a Fully convolutional networks (FCN) *Dive into Deep Learning* https://d2l.ai/chapter_computer-vision/fcn.html

Zhang J, Tian X, Shao J, Luo H and Liang R 2019b Phase unwrapping in optical metrology via denoised and convolutional segmentation networks *Opt. Express* **27** 14903

Zhang Z, Dai Y and Sun J 2020 Deep learning based point cloud registration: an overview *Virtual Real. Intell. Hardw.* **2** 3–222

Zhao Z, Cheng Y, Shi X and Qin X 2019 Classification method of LiDAR point cloud based on three-dimensional convolutional neural network *J. Phys. Conf. Ser.* **6** 1168

Zhong C, Gao Z, Wang X, Shao S and Gao C 2019 Structured light three-dimensional measurement based on machine learning *Sensors* **19** 3229

IOP Publishing

Advances in Optical Form and Coordinate Metrology

Richard Leach

Chapter 7

Precision freeform metrology

Chi Fai Cheung, Lingbao Kong and Mingjun Ren

Freeform surfaces are widely used in many fields ranging from the design and manufacturing of precision dies, moulds and products in the plastics, automotive and aerospace industries, to biomedical, entertainment, bionics and geographical data processing applications. Although it is now feasible to fabricate ultra-precision freeform surfaces with sub-micrometre form accuracy and surface finish in the nanometre range, there is currently a lack of international specification standards governing the measurement of freeform surfaces. Some equipment manufacturers have attempted to deliver ultra-precision freeform measurement instruments to the market. However, the understanding of key issues and challenges in precision freeform metrology is still far from complete. In this chapter, the key issues, challenges and state-of-the-art technologies and methods for the measurement and characterisation of ultra-precision freeform surfaces are reviewed and discussed, to shed some light on the possible future developments towards advancement and standardisation.

7.1 Overview of freeform surfaces

Unlike conventional surfaces, such as planes, spheres and cylinders that usually possess rotational and translational symmetry, by definition, freeform surfaces are surfaces without rotational or translational invariance (ISO 2011). Structured surfaces, such as vee-grooves, microlens arrays and micro-pyramids, are also considered freeform surfaces (Jiang *et al* 2007a). Compared to a simple surface, freeform surfaces often provide superior functional performance (Jiang *et al* 2007b) and are now widely used in many industries, such as aerospace, automotive, optics, biomedical and telecommunications.

As an example, one of the most important attributes of freeform surfaces is their ability to affect the aerodynamic performance for moving objects such as airplanes, automobiles and wind turbine blades. With careful design and computational simulation, together with the technologies for manufacturing and measurement of precision freeform surfaces, the drag coefficient can be significantly reduced, since the flow condition can be improved with purposely designed continuous freeform surfaces. For automobiles, 40% of the drag coefficient can be attributed to the shape of the vehicle (Hucho and Sovran 1993) and current wind turbine blades make use of freeform design to increase their aerodynamic efficiency.

Freeform surfaces are widely used in biomedical implant applications, since most of the parts of the human body are freeform shapes (Curodeau *et al* 2000). For example, the replacement of hips and knees in biomedical applications requires the polishing and correction of freeform surfaces (Charlton and Blunt 2008). The surface accuracy and surface finishing of these implants impose stringent requirements and the implants are usually made of difficult-to-machine materials for their bio-capability. Figure 7.1(a) shows an implant used for a total knee replacement.

Freeform surfaces are widely used in advanced optics, since optical performance can be significantly improved and the size of the optical system can be reduced. Applications of freeform optics are also found in the optometric field. For example, the freeform progressive addition lens (PAL) contains a far view zone, a near zone and a progressive zone between the far and near zones (Kong *et al* 2014). Figure 7.1(b) shows a machined PAL mould insert with a freeform surface. Freeform optics are also widely used in illumination systems to improve the efficiency of light transfer or reduce the number of optical elements (Jiang *et al* 2006). Freeform lenses are widely used in lighting systems, for example, LED lighting systems. Freeform lenses (see figures 7.1(c) and (d)) are used to redistribute

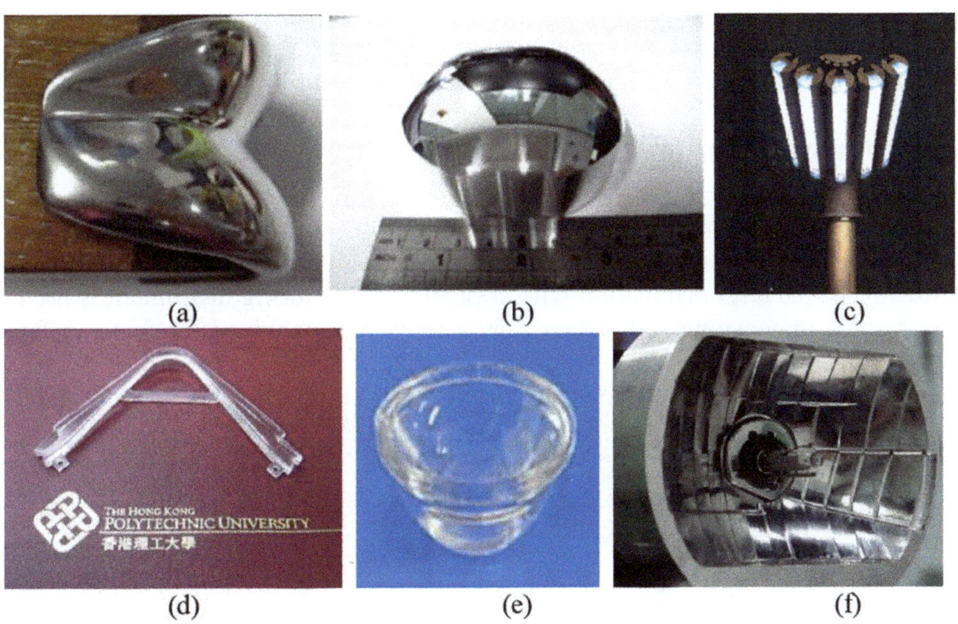

Figure 7.1. Examples of freeform workpieces: (a) biomedical implant; (b) freeform progressive lens mould insert; (c) LED street lamp; (d) optical light guide; (e) LED collimating lens; and (f) automotive headlamp reflector.

the light flux from the LED to meet the needs for lighting. Freeform surfaces are also used in optical light guides and automotive headlamp reflectors, as shown in figures 7.1(d) and (f), respectively.

Freeform surfaces with microstructures, such as microlens arrays (Cheung *et al* 2010), surfaces with Frustum ridge structures (Kong *et al* 2013), micro-grooves (Fang *et al* 2003) and micro-pyramids (Yan *et al* 2010), are widely used in advanced optics to improve performance and provide functionalities. Figure 7.2 shows a variety of freeform surfaces with microstructures. Microlens arrays have various applications for focusing, illuminating and imaging. The micro-ridge has self-cleaning and imaging functionalities and micro-grooves have a wide range of applications in the mechanical, electronic, photonic and biomedical fields.

Roll-to-roll manufacturing of microstructures is a promising technique to achieve mass production. There are many freeform applications in the roll-to-roll industry for mass production of everyday-life products incorporating microstructures. Since the dimensional scale of the microstructures is small—several micrometres or even smaller—the manufacturing and measurement of these features is challenging. Figure 7.3 shows an example of a roll-to-roll process and the machined vee-groove film. The vee-groove was fabricated on a 100 μm thick film substrate.

With the advancement of ultra-precision machining technology, it is now technically feasible to fabricate ultra-precision freeform and structured surfaces with sub-micrometre form accuracy and surface finish in the nanometre range. However, due to the geometrical complexity and high precision requirements, there is still a lack of international specification standards to support traceable measurement of freeform and structured surfaces, so as to provide governance of quality assurance in their precision manufacturing. In this chapter the state-of-the-art technology for precision freeform metrology is discussed.

7.2 Framework for precision freeform metrology

Figure 7.4 shows a framework for precision freeform metrology. In general, the surface measurement and form characterisation of ultra-precision freeform surfaces can be divided into two parts: data acquisition and data processing, as shown in figure 7.5. In the data acquisition part, a precision measuring instrument is used to extract three-dimensional (3D) data points from the machined freeform surfaces.

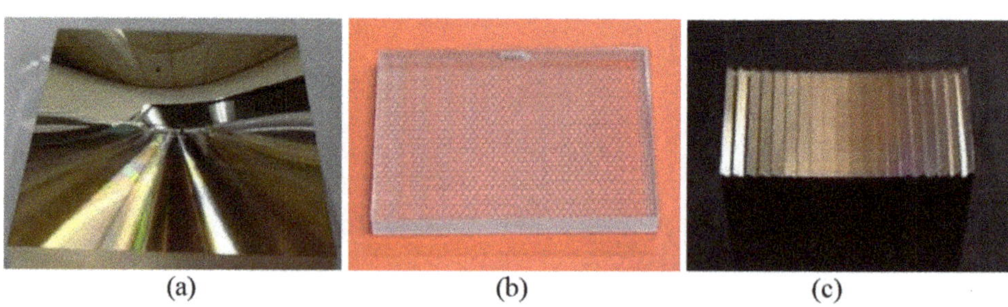

Figure 7.2. Freeform surfaces with microstructures: (a) Fresnel lens mould insert; (b) light guide with a microlens array on one side and a vee-groove on the other side; and (c) inert vee-groove mould.

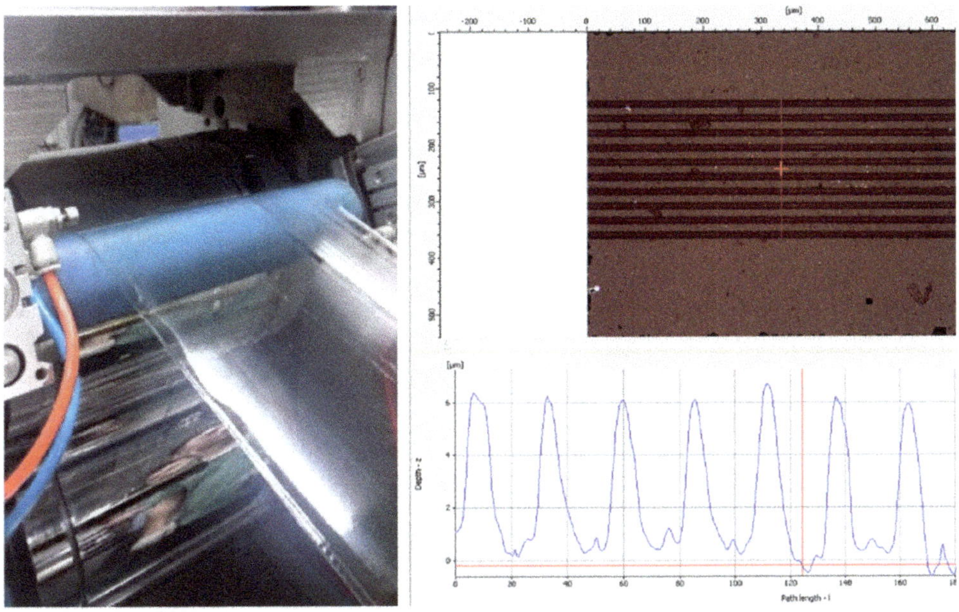

(a) (b)

Figure 7.3. The roll-to-roll fabrication process. (a) Roll-to-roll equipment. (b) Film with vee-groove microstructures.

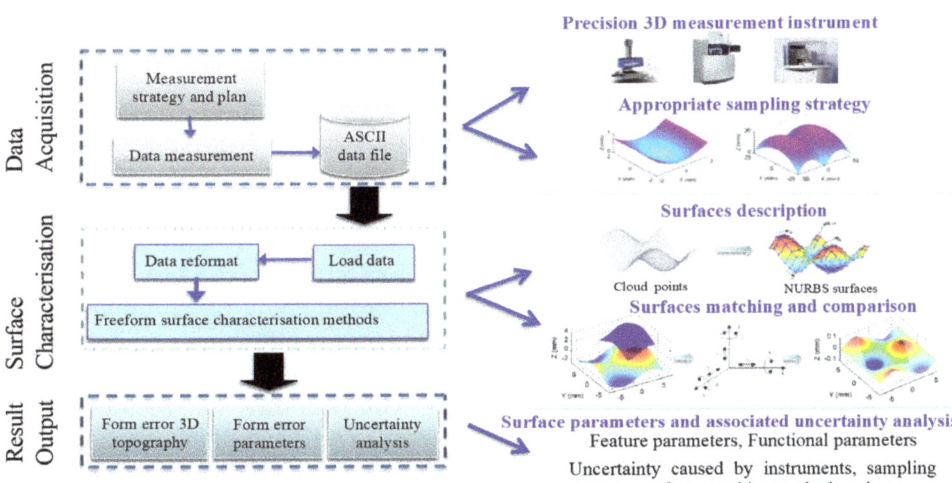

Figure 7.4. Framework for precision freeform metrology.

An appropriate sampling strategy is critical to ensure that the number of measured data points is adequate to fully represent the geometry of the machined surface.

In data processing, form characterisation methods are used to evaluate the form error of the measured surface (MS) by comparing with that of the design surface

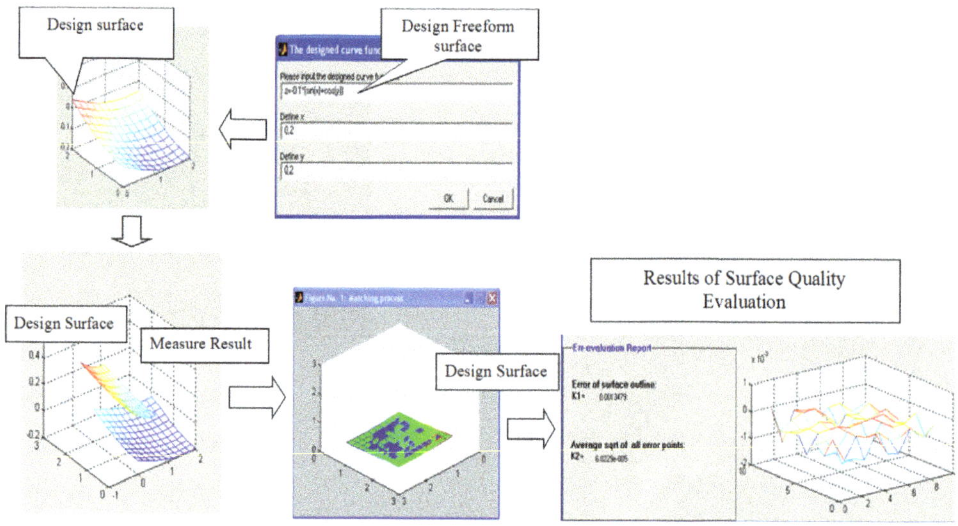

Figure 7.5. Graphical illustration of freeform surface characterisation.

(DS). Surface matching or correspondence searching between the MS and the DS is an important step, since the measured data points are embedded in the coordinate system of the corresponding measuring instrument, which is commonly not the same as the coordinate system of the DS. Small misalignments between the two coordinate systems can cause uncertainty in the form error evaluation results. Currently, least-squares methods are the most widely used in freeform surface matching. However, any measurement process has errors which can lead to uncertainty in the measurement result. The uncertainty associated with the measurement and form characterisation of freeform surfaces comes from many sources.

In data acquisition, the uncertainty can be due to the measuring instrument, including the uncorrected systematic and random probing errors, probing parameters and the dynamics of the instrument mechanical loops. The sampling strategy is also a significant contributor to the measurement uncertainty (Hocken 1993)—there can be errors due to inadequate sampling and the sampling distribution. In data processing, the uncertainty is dominated by the surface matching and comparison methods used. Since the measured data do not perfectly match with the DS, the surface matching results always contain errors. Although there has been research in recent years to develop more effective and accurate surface matching, from coarse to fine matching (Sharp *et al* 2002, Yamany and Farag 2002, Li and Gu 2005, Jiang *et al* 2010), the associated uncertainty in the characterisation results is rarely quoted.

Research on the analysis of the uncertainty in data processing in geometric measurement is often focused on simple geometries, such as circles, spheres and cylinders (Maihle *et al* 2009a, 2009b, Zhang *et al* 2006), and there has been relatively little research for freeform surfaces. Some researchers attempt to represent freeform surfaces by breaking them down into simple features, so that the problem is converted to the analysis of simple geometries (Savio and De Chiffre 2002), but this method cannot be applied for many kinds of freeform surfaces. The general model described by the *Guide to the Expression of Uncertainty in Measurement* (GUM) (JCGM 100 2008) is difficult to apply to freeform surface measurement, since the uncertainty varies with the nature of the surfaces being measured. As a result, a task-specific uncertainty evaluation method is often needed, for example, the associated uncertainty in form characterisation results has been evaluated when the measured data are extracted from a specific surface (Ren *et al* 2012), including measurement error, surface form error and sampling strategy.

7.3 Characterisation of freeform surfaces

This section discusses the post-processing technology for the measured data which are acquired as coordinate data (points clouds) from various measuring instruments, such as contact and optical coordinate measuring systems (CMSs), form/profile measuring instruments, interferometers and other methods (see Leach and Shaheen (2020) for a recent review of instruments). Figure 7.5 shows a graphical illustration of the process for form characterisation of a freeform surface. This process usually includes surface description and presentation by surface fitting and reconstruction, surface matching and evaluation of form errors with surface parameters. Surface fitting aims to generate a continuous surface to carry out the surface matching process when the MS and DS data are discrete point clouds. Surface matching is necessary when the freeform surface is characterised by comparing the MS data with the DS data. Finally, surface parameters are used to present the results of surface characterisation.

7.3.1 Surface fitting and reconstruction

Freeform surfaces do not have rotational symmetry and cannot be generalised by equations used for rotational symmetric surfaces, such as aspheric surfaces. The representation of freeform surfaces is usually based either on a known surface model or a cloud of discrete measured data for an unknown surface model (Cheung *et al* 2006). If the surface model is not available and only a cloud of discrete measured data of the surface is provided, the unknown surface model must first be reconstructed in order to obtain the form of the theoretical surface from the discrete data points.

In the measurement of freeform surfaces, a large number of discrete data points are normally acquired to fully describe the geometry of the surface. However, performing high-density measurement of freeform surfaces can be time consuming and costly for contact CMS or point-scanning optical CMS. As shown in figure 7.6, a certain number of discrete measured data points are extracted from the machined surface, using an appropriate measurement strategy, and the measured data are fitted by a continuous surface to represent the MS, including non-measured areas, and for further mathematical processing in the form characterisation (Savio *et al* 2007). This process infers that surface fitting and reconstruction are critical in the measurement and form characterisation so as to represent the surface shape based on a cloud of discrete measured points from the workpiece.

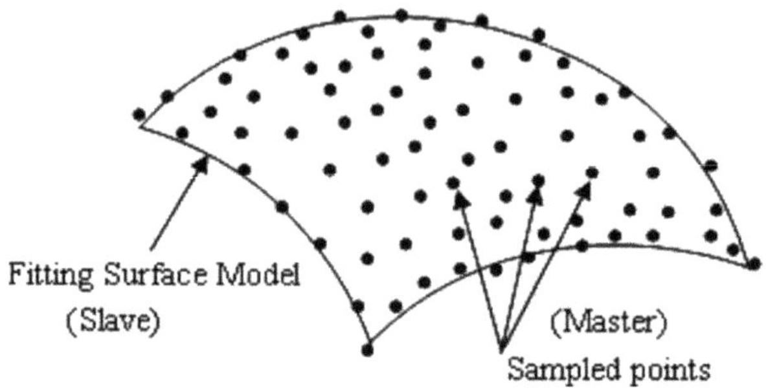

Figure 7.6. Measured data points and the fitting surface model (Cheung *et al* 2009). Copyright IOP Publishing. Reproduced with permission. All rights reserved.

Due to their high degree of geometrical freedom and relatively simple mathematical structure, B-spline surfaces are commonly adopted for constructing freeform surfaces. A B-spline surface $\boldsymbol{S}$ is defined as (Les and Wayne 1997):

$$\boldsymbol{S}(u, v)|_{p,q} = \sum_{i=1}^{i=n_u}\sum_{j=1}^{j=n_v} N_{i,p}(u)N_{j,q}(v)\boldsymbol{P}_{ij} \quad 0 \leqslant u \leqslant 1, 0 \leqslant v \leqslant 1, \tag{7.1}$$

where (p, q) are the degree of the B-spline surface; $\boldsymbol{P}_{ij}$ is the point controlling the shape of the surface; n_u and n_v are the numbers of control points in the u and v directions, respectively; u and v are surface parameters identifying the location of point $\boldsymbol{S}(u, v)$ within the length of the two directions of the surface; $N_{i,p}(u)$ and $N_{j,q}(v)$ are the normalised B-spline functions uniquely defined by the degree p and a knot vector U, degree q and a knot vector V, respectively. Here, the $(p = 3, q = 3)$th degree of the B-spline surface is adopted. The surface reconstruction can then be formulated as an optimisation scheme (Weiss *et al* 2002): given a set of discrete data points $\boldsymbol{X}_k$, $k = 1, 2, \ldots m$, a B-spline surface $\boldsymbol{S}(u, v)$ can be constructed with a set of control points $\boldsymbol{P}_{ij}$, $i = 1, 2, \ldots n_u$, $j = 1, 2, \ldots n_v$ and appropriate knot vectors, such that the objective function is given by

$$F = \frac{1}{2}\sum_{k=1}^{m} d^2\left(\sum_{i=1}^{n_u}\sum_{j=1}^{n_v} N_{i,3}(u)N_{j,3}(v)\boldsymbol{P}_{ij}, \boldsymbol{X}_k\right) + \lambda_s f_s. \tag{7.2}$$

This objective function is minimised by satisfying the constraint $\min_{(u, v)} \| \boldsymbol{S}(u, v) - \boldsymbol{X}_k \| \leqslant \varepsilon_f$, where $d(\boldsymbol{S}(u, v), \boldsymbol{X}_k)$ is a distance function that is used to represent the approximation error of the reconstructed surface from $\boldsymbol{X}_k$; f_s is a smoothing term, λ_s is the weight of f_s and ε_f is a user-defined tolerance.

This optimisation scheme is a highly non-linear problem and is difficult to solve, in particular when the required tolerance for approximation is tight (Franke and Nielson 1991, Dierckx 1995, Weiss *et al* 2002). If the discrete points are given as a regular lattice (points on a grid), the problem can be solved by a tensor product B-spline surface using chord-length parametrised knot vectors. However, if the points are given as an unorganised points cloud, the surface is generally reconstructed in two steps. An initial surface is first constructed to approximate the point cloud with a defined precision (Dierckx 1995, Weiss *et al* 2002). This initial surface is used to estimate the minimal degree of freedom needed to characterise the real shape, and to obtain a good parametric value for each corresponding data point set by projecting the point cloud onto it.

There are various methods to construct the initial surface. Ma and Kruth (1995) used interactively defined section curves together with the four boundary curves to obtain an initial surface. Weiss *et al* (2002) proposed a recursive method to construct the initial surface, starting with a simple surface with a few control points and gradually increasing the smoothness or the number of control points until the control distribution of points of the initial surface is satisfactory. The constructed initial surface is then used to assign parameters to all fitted points, i.e. locate the points on the initial surface that are close to the fitted points. Approximations of higher order or accurate foot point computations are discussed elsewhere (Hoschek and Lasser 1993, Saux and Daniel 2003, Hu and Wallner 2005).

The initial surface is optimised to meet the prescribed tolerance. Once each data point is assigned corresponding parameters, the distance function from the fitted data to the surface can be formulated to optimise the initial surface. Wang *et al* (2006) gave an overview of fitting methods and divided them into three categories based on different distance functions: point distance minimisation (PDM), tangent distance minimisation (TDM) and squared distance minimisation (SDM). PDM minimises the squared distance from the data points to the foot points to optimise the constructed surface. PDM is the most popular technique for surface fitting in computer graphics and computer aided design (CAD) with B-spline surfaces, as well as other types of surface, due to its simplicity (Hoppe *et al* 1994, Farin 1996, Hoppe 1998, Haber *et al* 2001, Greiner *et al* 2002).

Blake and Isard (1998) used TDM from the data point to the tangent plane in the foot point as a basic error term to speed up the convergence of the iterative optimisation. SDM was studied by Pottmann and Hofer (2003) and was used to solve curve and surface fitting problems (Pottmann *et al* 2002). Wang *et al* (2006) proposed an algorithm based on a second order approximation to the distance function (SDM), to solve the curve fitting problem. They compared the performance of SDM with those of PDM and TDM in fitting a B-spline curve to point clouds and found that SDM converges faster than PDM, and is more stable than TDM.

Multilevel B-spline approximation of 3D discrete points has been studied by Lee *et al* (1997, 2005). This approximation makes use of a coarse-to-fine hierarchy of control lattices to iteratively perform the local refinement, such that a B-spline surface is reconstructed with high fidelity. In addition to minimising the approximation error of the reconstructed surfaces, various smoothness functions are usually added to improve the quality of the reconstructed surface. The most frequently used function is the simplified quadratic function of the parametric derivatives (Dietz 1998). Traditional fitting technologies can efficiently reconstruct smooth surfaces from a cloud of discrete points, but few of them can clearly demonstrate how smooth the reconstructed surfaces must be so as to represent the real form of the surface. It is critical to specify an appropriate fitting threshold to balance the fitness and smoothness of the reconstructed surface. This is particularly true for the reconstruction of a measured surface from a cloud of measured data. Moreover, most of the existing optimisation methods for the initial surface are based on local optimisation, which makes the optimised surfaces highly sensitive to the initial surface (Yang *et al* 2004). It is also critical to determine the appropriate number of control points and their distribution on the initial surface to ensure the accuracy of the optimised surface in meeting a prescribed error threshold.

Knots that are set evenly in space may cause overshoot problems in the case of smooth uneven or discontinuous data. To overcome this problem, the determination of an uneven knotting space can be applied. Due to the degree of computation, defining knots and their corresponding locations in non-uniform spaces remains a challenging problem. Data characteristics are an effective basis for determining the location of non-uniform knots. Variable coefficients based on diagonalisable homogeneous linear differential systems have explicit solutions, which can be generalised to surface fitting with reasonable accuracy (Li *et al* 2018). The method of semi-segmentation and control with a pre-determined threshold is used to find the best-fitting piecewise linear function on the calculated second derivative (Tjahjowidodo *et al* 2015). Bicubic B-spline surfaces constrained by partial differential equations can fully determine internal control points by giving control points for the boundaries of open or closed surfaces (Han and Han 2019). The multi-feature fitting algorithm combines the least-squares method and the density-based outlier detection method (Liu *et al* 2018), and the two methods run alternately to gradually achieve the best fit results and eliminate interference caused by outliers. The multi-feature fitting method based on data features is simple and fast, but it is not robust to non-contained features.

Local search methods can significantly improve the efficiency of non-uniform knot setting (Iglesias *et al* 2016). A non-uniform interpolation subdivision scheme generalises the uniform interpolation four-point scheme to arbitrary topological grids (Li and Chang 2018). The point projection method can be used to improve the accuracy of the multilevel B-spline approximation (MBA) method (Moon and Ko 2018). By directly adjusting the control points of the local area and combining it with the MBA method, the approximation error can be reduced. Typical global optimisation methods can also be used to automatically adjust the knots of the B-spline curve, and automatically determine the number and position of the knots to obtain an accurate data fit, such as the adaptive clone selection algorithm (Gálvez *et al* 2015), the improved genetic algorithm (Sui *et al* 2016) and multi-objective particle swarm optimisation (Hashemian and Hosseini 2018).

New surface reconstruction techniques can effectively improve the accuracy of the fitting, such as implicit B-spline (Rouhani *et al* 2014), omega B-spline curves (Tang *et al* 2018), gradient-based fitting for knots (Ćurković *et al* 2018) and hierarchical B-splines (Garau and Vázquez 2018). By changing the construction of the generalised B-spline curve basis, a better approximation can be obtained without considering the orthogonal basis of the corresponding space, such as constructing a double generalised B-spline function (Zhang *et al* 2015).

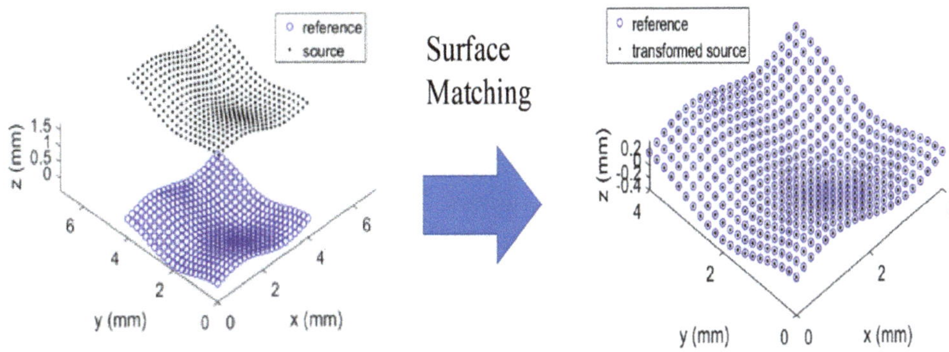

Figure 7.7. Graphical illustration of surface matching for freeform surfaces.

7.3.2 Surface matching

The form error of an MS is evaluated by comparing it with the DS. One of the key issues is surface matching, referred to as surface fitting by Jiang *et al* (2010) and localisation/registration by Li and Gu (2004) between the MS and the design template (Li and Gu 2004, Savio *et al* 2007, Cheung *et al* 2009). The reason for surface matching is that the measured data points are embedded in the coordinate system of the corresponding measuring instrument, which is usually not the same as the coordinate system of the design template. Small misalignments between the two coordinate systems can cause significant errors in the results of the form error evaluation. This issue is critical for the measurement of ultra-precision freeform surfaces, which have form accuracies down to sub-micrometre levels. Hence, the measured data and the designed template have to be accurately transferred to a common coordinate system before the comparison, as illustrated in figure 7.7.

Form characterisation of freeform surfaces is commonly formulated as an optimisation problem to search for an optimal Euclidean motion (translation and rotation) based on methods, such as least-squares, as represented in equation (7.3) or the minimum zone method, as represented in equation (7.4), so that the two surfaces are aligned as closely as possible (Besl and McKay 1992, Menq *et al* 1992, Yau and Menq 1996):

$$F = \sum_{i=1}^{n} | \mathbf{P}_i - \boldsymbol{T}\mathbf{Q}_i |^2 \tag{7.3}$$

$$F = \min(\max | \mathbf{P}_i - \boldsymbol{T}\mathbf{Q}_i | - \min | \mathbf{P}_i - \boldsymbol{T}\mathbf{Q}_i |) \qquad i = 1, \ldots n, \tag{7.4}$$

where $\mathbf{Q}_i$ is the measured point and $\mathbf{P}_i$ is the corresponding point of $\mathbf{Q}_i$ in the template surface; $| \cdot |$ is the distance between $(\mathbf{P}_i, \mathbf{Q}_i)$ and is positive if $\mathbf{Q}_i$ is above the template surface, otherwise negative. $| \mathbf{P}_i - \boldsymbol{T}\mathbf{Q}_i |$ is the deviation of the measured surface at point $\mathbf{Q}_i$. The coordinate transformation $\boldsymbol{T}$ is a function of spatial rotation and translation and can be represented by

$$\boldsymbol{T} = \begin{bmatrix} c(r_z)c(r_y) & s(r_z)c(r_y) + c(r_z)s(r_y)s(r_x) & s(r_z)s(r_x) - c(r_z)s(r_y)c(r_x) & t_x \\ -s(r_z)c(r_y) & c(r_z)c(r_x) - s(r_z)s(r_y)s(r_x) & c(r_z)s(r_x) + s(r_z)s(r_y)c(r_x) & t_y \\ s(r_y) & -c(r_y)s(r_x) & c(r_y)c(r_x) & t_z \\ 0 & 0 & 0 & 1 \end{bmatrix}, \tag{7.5}$$

where t_x, t_y and t_z are the translation distances, and r_x, r_y and r_z are the rotation angles along the $\vec{X}$-, $\vec{Y}$- and $\vec{Z}$-axis, as shown in figure 7.8; and $c()$ and $s()$ are abbreviations for the cosine and sine functions, respectively. The core of this optimisation scheme is the establishment of correspondence pairs ($\mathbf{P}_i$, $\mathbf{Q}_i$) between two matching objects. Using a six degrees-of-freedom rigid body transformation, the MS is matched to the DS model with the modification of positions (t_x, t_y, t_z) and attitudes (r_x, r_y, r_z). This is a multi-step optimisation process. For each alignment step, the distance from the measured point to the corresponding point in the DS needs to be calculated as shown in figure 7.9.

In measurement, neither the corresponding points of the measured data, nor the coordinate transformation function $\boldsymbol{T}$ are known. Hence, the freeform surface matching problem is generally solved with a nested approach (Besl and McKay 1992, Ahn 2004). First, the correspondence of each measured point is established by locating the minimum distance point on the DS in the inner iteration. In the outer iteration, $\boldsymbol{T}$ is estimated based on the established correspondence pairs. $\boldsymbol{T}$ is then used to transform the MS and new correspondence pairs are established. This process is iterated and is terminated when the desired accuracy is achieved.

Due to the non-convexity of the optimisation problem, the matching results may be trapped at a local minimum or even become divergent if the initial relative position of the two matching surfaces is not appropriately provided (Ahn 2004). Hence, conventional methods perform freeform surface matching in two stages: coarse matching and fine matching (Li and Gu 2004, Zhang 2009). Coarse matching is intended to find a rough position for the MS with respect to the DS. The rough matching results are optimised at the fine matching stage.

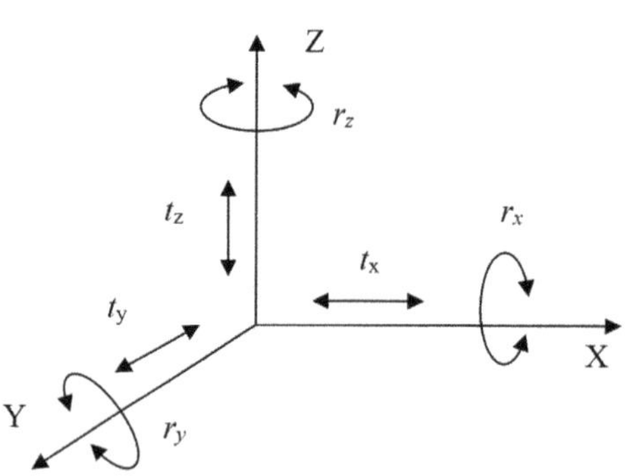

Figure 7.8. Six degrees of rigid body transformation (adapted from Cheung *et al* 2009). Copyright IOP Publishing. Reproduced with permission. All rights reserved.

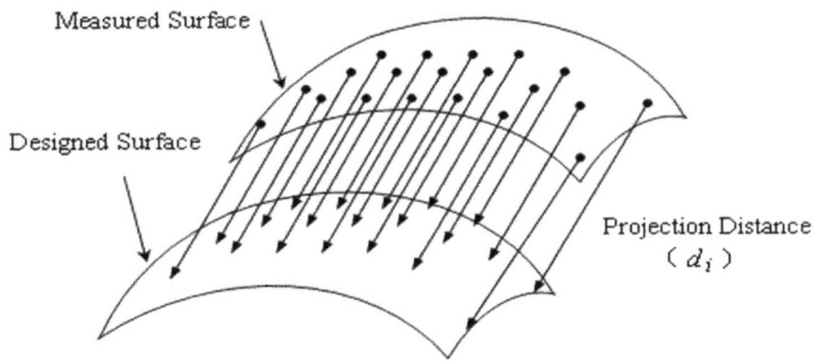

Figure 7.9. Matching by rigid body transformation from the measured surface to the designed one (Cheung *et al* 2009). Copyright IOP Publishing. Reproduced with permission. All rights reserved.

7.3.2.1 Coarse matching of freeform surfaces

If the measured surface includes some salient features, such as holes, slots or pockets, these features can be used to perform coarse matching (Joshi and Chang 1988). However, most smooth freeform surfaces do not contain any salient features. As a result, various orientation independent mathematical and geometrical surface features have been developed. Cheung *et al* (2006) developed a method called the five-point method. The gravity centre and the four corner points of the surface are used to construct a feature shape, which is used to perform surface matching. These methods are effective when the MS and the DS are approximately of the same size. Corney *et al* (2002) proposed a method to calculate the convex hull of a 3D object. Values such as hull 'crumpliness', hull packing and hull compactness are obtained from the convex hull and are taken as measures of the similarity between two objects.

Ko *et al* (2003a, 2003b) utilised the Gaussian and mean curvatures for surface matching, and the related iso-curvature lines were used to establish the correspondence between two objects. Other surface intrinsic properties of umbilical points and intrinsic frames were also used to carry out freeform object matching. Li and Gu (2005) defined four types of surface shape: concave surfaces, convex surfaces, saddle surfaces and flat surfaces based on different geometric information about the surfaces. The rough matching was carried out by searching four correspondences, including feature similarity, inter-feature distance, curvature similarity and frame similarity, and the fine localisation was conducted by the ICP method (see chapter 2).

Zhu *et al* (2009) proposed a 3D shape matching approach based on diffusion wavelets. Jiang *et al* (2010) proposed a new method for freeform surface matching whereby a new surface feature named the structured region signature was used to conduct rough matching. Cheung *et al* (2006, 2009) and Kong *et al* (2007) presented a series of surface matching algorithms for optical freeform surfaces. They presented several freeform surface coarse matching methods, including the five-point method, coupled reference data method, and a hybrid surface fitting and matching method.

Some global features, such as geometric moments (Funkhouser *et al* 2003), Fourier descriptors (Zhang and Lu 2002) and harmonic shape images (Groemer 1996), can be used in 3D surface matching. Novotni and Klein (2004) proposed a 3D Zernike descriptor for 3D shape searching. Zernike moments are considered as having optimal performance in terms of noise and information redundancy. Guo and Li (2009) presented a new shape descriptor, named Gaussian curvature moment, invariant for surface matching. It was claimed that the new descriptor was superior in accuracy and efficiency to Fourier descriptors and Zernike moments.

Redefining point set registration to a point-to-model alignment problem with implicit surface fitting can avoid time-consuming corresponding searches. However, under the rigid transformation, the minimum value of the sum of squares of the approximate distance between the point set and the model easily falls into the wrong position. Non-linear least-squares optimisation of rigid transformation parameters at equivalent distances can converge to the optimal position with a wider convergence range and lead to more accurate transformation parameters (Liu *et al* 2015). The enhanced Hough transform and the improved RANSAC transform method (see chapter 2) introduce curvature parameters into the algorithm for coarse alignment to address the limitations of principal component analysis (PCA) (Rantoson *et al* 2016). These curvature parameters are pre-determined for each vertex of the mesh data through a controlled mesh discretisation. The local transformation calculations given by Huff's exhaustive search apply only to points of the same surface type. The random sampling of the points to be paired and their correspondence search in the RANSAC-based dance algorithm focus on the reduced area specified by a mixture of surface types and curvature similarity constraints.

7.3.2.2 Fine matching of freeform surfaces

The rough position obtained by coarse matching is refined by fine matching, the most common of which is ICP (Besl and McKay 1992, chapter 2). ICP is an iterative descent procedure which seeks to minimise the sum of the squared distances between all points in MS and their closest points in the corresponding DS. When the MS and the DS are represented as two sets of data points with known correspondences, the coordinate transformation matrix can be determined by solving equation (7.3) or (7.4).

In the iterative optimisation process, the dynamic weighting strategy controls the priority of different surfaces in the positioning process and makes the positioning beneficial to surfaces with higher quality requirements (Tan *et al* 2015). Since no additional constraints are required for different accuracy requirements, the proposed differential positioning is an unconstrained non-linear optimisation problem. The improved Newton iteration fitting algorithm can generate initial values and improve the fitting speed (Zhang *et al* 2017). Dynamic PCA (DPCA) uses PCA of the measurement data and the information generated by the corresponding fitting geometry to fit complex surfaces to coordinate metrology measurement points (Gohari and Barari 2016).

Although many kinds of orientation invariant surface descriptors have been developed to perform surface matching, many of them are either susceptible to noise and outliers in the measured data or the features become inconspicuous when the surface to be matched is relatively smooth and flat. These issues can introduce

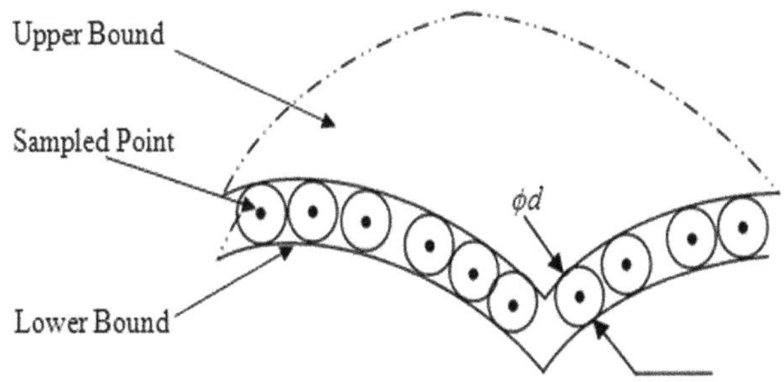

Figure 7.10. Tolerance zone of a freeform surface.

uncertainty to the surface matching results. Despite research to improve computation efficiency, surface matching under conventional schemes still involves extensive computation. In each iteration, it requires the solution of highly non-linear equations (normally also an iterative process), and establishing correspondence pairs by searching the closest point for each measured point in the DS. In practice, the characterisation of ultra-precision freeform surfaces usually involves a large number of measured data points, which often leads to unacceptable computing times.

7.3.3 Surface parameters for form error evaluation

The manufacturing specification of freeform surfaces is commonly described by form tolerance, which is the permitted maximum value of the form deviation. According to ISO 1101 (ISO 2017), there is a defined form tolerance zone within which all points of the features must be contained, and the machined surface can be of any form if it is not specified. Figure 7.10 shows the construction of a tolerance zone for a freeform surface. A set of points are sampled from the DS; a series of spheres are established with each sampled point as the centre point at a diameter d. Two envelope surfaces are constructed based on these spheres and denoted as the upper bound surface and the lower bound surface. The zone between the upper and lower surfaces is the tolerance zone of the DS with tolerance d.

For conventional surfaces, such as planes, spheres and cylinders, standard parameters, such as flatness, straightness, cylindricity and circularity, can be used to characterise the form accuracy of the workpiece. Due to the geometric complexity of freeform surfaces, there are currently no standard surface parameters for form characterisation (Jiang *et al* 2007b, Whitehouse 2011). The form error of a freeform surface at a measured point is determined by projecting this point onto corresponding design template after surface matching. The form error of a measured freeform surface is characterised by areal surface parameters (ISO 2017) which are determined by the local form error of the measured points.

Areal surface parameters have been developed as an extension of surface characterisation from 1D profiles to 2D surfaces (see Leach 2014 for a detailed treatment of areal parameters). Height parameters are amplitude parameters which are defined with respect to a mean plane obtained through the levelling of the mean square plane of the measured surface. The height parameters can be used to characterise the form errors of measured freeform surfaces.

Other groups of parameters include spatial parameters, feature parameters, functional parameters and hybrid parameters (Leach 2014) and these have been used to characterise freeform products for some specific applications. Sequential quadratic programming (SQP)-based algorithms simplify the process of updating the alignment parameters through an approximately linear differential motion model with a signed distance (Lang *et al* 2017). The surface subdivision method or stereo lithography (STL) model is used to determine the point-to-surface distance to evaluate the contour error. Key points extracted based on wavelet decomposition have been used to locate the key point set by using SQP algorithms and determining the contour error (He *et al* 2019). To simplify the calculation and maintain the accuracy of the evaluation at the same time, those measurement points that can be defined by a significant deviation from the DS could be used to evaluate the contour error of the freeform surface according to the decomposition results.

7.4 Conclusions and future research

Freeform surfaces are used in the development of a wide range of products to realise specially designed optical and mechanical functions. The geometry of these surfaces must be accurate to ensure their functional performance. However, the complexity of ultra-precision freeform surfaces creates considerable difficulty for the control of the manufacturing process and quality assessment. This chapter provides an overview of current precision freeform metrology, focusing on the surface characterisation. Surface characterisation plays an important role for quantifying the form error of the measured surface through surface fitting and reconstruction, and rough and fine matching of the design surface and measured surface. The errors of the measured surface are then quantified by surface parameters.

In the last decades, freeform surface characterisation methods have received a great deal of research attention. It is interesting to note that most methods are based on surface matching of distance errors. Many researchers employ the least-squares or minimum zone methods as the optimisation criteria for matching the design and measured surfaces. Due to the geometrical complexity of freeform surfaces, this create some challenges in freeform surface characterisation including uncertainties in surface fitting and reconstruction as well as surface matching. There is also a lack of standardisation of surface parameters for freeform surface characterisation.

There is a need to develop new methods to perform high-precision and robust characterisation of the form and surface texture of freeform surfaces. A versatile precision metrology platform with high accuracy and robust traceability is much needed to meet the increasing demand for measuring and evaluating the surface quality for freeform components. Such methods would help to achieve robust and traceable measurement and validation of the surface quality for freeform surfaces. Moreover, the development of surface parameters which can better correlate the surface feature to the functional performance of the freeform surface would be indispensable for the control of the manufacturing process and quality assessment.

Acknowledgments

The work described in this chapter was supported by grants from the Ministry of Science and Technology of China (Project Codes: 2017YFA0701200 and 2017YFE0191300). The authors would also like to express their thanks to the Research Office of The Hong Kong Polytechnic University (Project code: BBX7).

References

Ahn S J 2004 *Least Squares Orthogonal Distance Fitting of Curves and Surfaces in Space* (Berlin: Springer)

Besl P J and McKay N D 1992 Method for registration of 3-D shapes *Proc. SPIE* **1611** 586–606

Blake A and Isard M 1998 *Active Contours* (New York: Springer)

Charlton P and Blunt L 2008 Surface and form metrology of polished 'freeform' biological surfaces *Wear* **264** 394–9

Cheung C F, Hu K, Jiang X Q and Kong L B 2010 Characterization of surface defects in fast tool servo machining of microlens array using a pattern recognition and analysis method *Measurement* **43** 1240–9

Cheung C F, Kong L B, To S and Lee W B 2009 Measuring ultra-precision freeform surfaces using a hybrid fitting and matching method *Meas. Sci. Technol.* **20** 105103

Cheung C F, Li H, Kong L B, Lee W B and To S 2006 Measuring ultra-precision freeform surfaces using a robust form characterization method *Meas. Sci. Technol.* **17** 488–94

Corney J, Rea H, Clark D, Pritchard J, Breaks M and MacLeod R 2002 Coarse filters for shape matching *IEEE Comput. Graph. Appl.* **22** 65–74

Ćurković M, Ćurković A and Vučina D 2018 Novel re-parameterization for shape optimization and comparison with knot-based gradient fitting method *Comput. Methods Appl. Mech. Eng.* **336** 304–32

Curodeau A, Sachs E and Caldarise S 2000 Design and fabrication of cast orthopedic implants with freeform surface textures from 3-D printed ceramic shell *J. Biomed. Mater. Res.* **53** 525–35

Dierckx P 1995 *Curve and Surface Fitting with Splines* (New York: Oxford University Press)

Dietz U 1998 Fair surface reconstruction from point clouds *Mathematical Methods for Curves and Surfaces II* ed M Dæhlen, T Lyche and L Schumaker (Nashville, TN: Vanderbilt University Press)

Fang F Z, Wu H, Liu X D, Lim G C, Liu Y C and Ng S T 2003 Fabrication of micro grooves *Proc. ASPE 18th Ann. Meeting (Portland, OR, USA, October)*

Farin G 1996 *Curves and Surfaces for Computer-Aided Geometric Design: A Practical Guide* (Amsterdam: Elsevier)

Franke R and Nielson G M 1991 Scattered data interpolation and applications: a tutorial and survey *Geometric Modeling* ed H Hagen and D Roller (Berlin: Springer)

Funkhouser T, Min P, Kazhdan M, Chen J, Halderman A, Dobkin D and Jacobs D 2003 A search engine for 3D models *ACM Trans. Graph.* **22** 83–105

Gálvez A, Iglesias A, Avila A, Otero C, Arias R and Manchado C 2015 Elitist clonal selection algorithm for optimal choice of free knots in B-spline data fitting *Appl. Soft Comput.* **26** 90–106

Garau E M and Vázquez R 2018 Algorithms for the implementation of adaptive isogeometric methods using hierarchical B-splines *Appl. Numer. Math.* **123** 58–87

Gohari H and Barari A 2016 A quick deviation zone fitting in coordinate metrology of NURBS surfaces using principle component analysis *Measurement* **92** 352–64

Greiner G, Kolb A and Riepl A 2002 Scattered data interpolation using data dependant optimization techniques *Graph. Models* **64** 1–18

Groemer H 1996 *Geometric Applications of Fourier Series and Spherical Harmonics* (Cambridge: Cambridge University Press)

Guo K and Li M 2009 A novel shape descriptor: Gaussian curvature moment invariants *1st Int. Conf. Inf. Sci. Eng. (Nanjing, China, December)* pp 1087–90

Haber J, Zeilfelder F, Davydov O and Seidel H P 2001 Smooth approximation and rendering of large scattered data sets *Proc. VIS'01 (San Diego, CA, USA, October)* pp 341–571

Han X and Han J 2019 Bicubic B-spline surfaces constrained by the biharmonic PDE *Appl. Math. Comput.* **361** 766–76

Hashemian A and Hosseini S F 2018 An integrated fitting and fairing approach for object reconstruction using smooth NURBS curves and surfaces *Comput. Math. Appl.* **76** 1555–75

He G, Sang Y, Wang H and Sun G 2019 A profile error evaluation method for freeform surface measured by sweep scanning on CMM *Precis. Eng.* **56** 280–92

Hocken R J 1993 Sampling issues in coordinate metrology *Manuf. Rev.* **6** 282–94

Hoppe H 1998 Efficient implementation of progressive meshes *Comput. Graph.* **22** 27–36

Hoppe H, DeRose T, Duchamp T, Halstead M, Jin H, McDonald J, Schweitzer J and Stuetzle W 1994 Piecewise smooth surface reconstruction *Proc. 21st Ann. Conf. Comput. Graph. Interact. Tech. (FL, USA, July)* pp 295–302

Hoschek J and Lasser D 1993 *Fundamentals of Computer Aided Geometric Design* (Natick, MA: A K Peters)

Hu S M and Wallner J 2005 A second order algorithm for orthogonal projection onto curves and surfaces *Comput. Aided Geom. Des.* **22** 251–60

Hucho W and Sovran G 1993 Aerodynamics of road vehicles *Ann. Rev. Fluid Mech.* **25** 485–537

Iglesias A, Gálvez A and Collantes M 2016 Four adaptive memetic bat algorithm schemes for Bézier curve parameterization *Transactions on Computational Science* vol 28 ed M L Gavrilova, C J K Tan and A Sourin (Berlin: Springer)

ISO 17450 part 1 2011 *Geometrical Product Specifications (GPS)—General Concepts—Part 1: Model for geometrical specification and verification* (Geneva: International Organization for Standardization)

ISO 1101 2017 *Geometrical Product Specifications (GPS)—Geometrical Tolerancing—Tolerances of Form, Orientation, Location and Run-Out* (Geneva: International Organization for Standardization)

JCGM 100 2008 *Evaluation of Measurement Data—Guide to the Expression of Uncertainty in Measurement* (Sèvres: Bureau International des Poids et Mesures)

Jiang J B, Cheung C F, Cheng S, Wang K W and Lee W B H 2006 Design and fabrication of freeform reflector for automotive headlamp *2nd Int. Conf. Power Electron. Syst. Appl. (Hong Kong, China, November)* pp 220–4

Jiang X, Scott P J, Whitehouse D J and Blunt L A 2007a Paradigm shifts in surface metrology. Part I. Historical philosophy *Proc. Math. Phys. Eng. Sci.* **463** 2049–70

Jiang X, Scott P J, Whitehouse D J and Blunt L A 2007b Paradigm shifts in surface metrology. Part II. The current shift *Proc. Math. Phys. Eng. Sci.* **463** 2071–99

Jiang X, Zhang X and Scott P J 2010 Template matching of freeform surfaces based on orthogonal distance fitting for precision metrology *Meas. Sci. Technol.* **21** 045101

Joshi S and Chang T C 1988 Graph-based heuristics for recognition of machined features from a 3D solid model *Comput. Aided Des.* **20** 58–66

Ko K H, Maekawa T and Patrikalakis N M 2003a An algorithm for optimal free-form object matching *Comput. Aided Des.* **35** 913–23

Ko K H, Maekawa T, Patrikalakis N M, Masuda H and Wolter F E 2003b Shape intrinsic properties for free-form object matching *J. Comput. Inf. Sci. Eng.* **3** 325–33

Kong L B, Cheung C F, To S, Lee W B and Cheng K W 2007 Measuring optical freeform surfaces using a coupled reference data method *Meas. Sci. Technol.* **18** 2252

Kong L B, Cheung C F, To S and Cheng C T 2013 Modeling and characterization of generation of 3D micro-structured surfaces with self-cleaning and optical functions *Optik* **124** 2848–53

Kong L B, Cheung C F, To S, Wang B and Ho L T 2014 A theoretical and experimental investigation of design and slow tool servo machining of freeform progressive addition lenses (PALs) for optometric applications *Int. J. Adv. Manuf. Technol.* **72** 33–40

Lang A, Song Z, He G and Sang Y 2017 Profile error evaluation of free-form surface using sequential quadratic programming algorithm *Precis. Eng.* **47** 344–52

Leach R K and Shaheen A 2020 Post-process metrology *Precision Metal Additive Manufacturing* ed R K Leach and S Carmignato (Boca Raton, FL: CRC Press)

Leach R K 2014 *Characterisation of Areal Surface Texture* (Berlin: Springer)

Lee S, Wolberg G and Shin S Y 1997 Scattered data interpolation with multilevel B-splines *IEEE Trans. Vis. Comput. Graph.* **3** 228–44

Lee W B, To S and Cheung C F 2005 *Design and Advanced Manufacturing Technology for Freeform Optics* (Hong Kong: The Hong Kong Polytechnic University)

Les P and Wayne T 1997 *The NURBS Book. Monographs in Visual Communication* (Berlin: Springer)

Li C J, Xie L L, Du W B, Li H D and Bao H 2018 Curve and surface fitting models based on the diagonalizable differential systems *Comput. Appl. Math.* **329** 179–90

Li X and Chang Y 2018 Non-uniform interpolatory subdivision surface *Appl. Math. Comput.* **324** 239–53

Li Y and Gu P 2004 Free-form surface inspection techniques state of the art review *Comput. Aided Des.* **36** 1395–417

Li Y and Gu P 2005 Feature-based alignment and comparison between portion and whole of free-form surfaces *Ann. CIRP* **54** 135–8

Liu T, Liu W, Qiao L, Luo T and Peng X 2015 Point set registration based on implicit surface fitting with equivalent distance *2015 IEEE ICIP (Quebec City, Canada, September)* pp 2680–4

Liu R, Wang Z and Liou F 2018 Multifeature-fitting and shape-adaption algorithm for component repair *J. Manuf. Sci. Eng.* **140** 021003

Ma W and Kruth J P 1995 Parameterization of randomly measured points for least squares fitting of B-spline curves and surfaces *Comput. Aided Des.* **27** 663–75

Maihle J, Linares J M and Sprauel J M 2009a The statistical gauge in geometrical verification. Part I. Field of probability of the presence of matter *Precis. Eng.* **33** 333–41

Maihle J, Linares J M and Sprauel J M 2009b The statistical gauge in geometrical verification. Part II. The virtual gauge and verification process *Precis. Eng.* **33** 342–52

Menq C H, Yau H T and Lai G Y 1992 Automated precision measurement of surface profile in CAD-directed inspection *IEEE Trans. Robot.* **8** 268–78

Moon S and Ko K 2018 A point projection approach for improving the accuracy of the multilevel B-spline approximation *J. Comput. Des. Eng.* **5** 173–9

Novotni M and Klein R 2004 Shape retrieval using 3D Zernike descriptors *Comput. Aided Des.* **36** 1047–62

Pottmann H, Leopoldseder S and Hofer M 2002 Approximation with active B-spline curves and surfaces *10th Pacific Conf. Comput. Graph. Appl. Proc. (Beijing, China, October)* pp 8–25

Pottmann H and Hofer M 2003 Geometry of the squared distance function to curves and surfaces *Visualization And Mathematics III* (Berlin: Springer)

Rantoson R, Nouira H, Anwer N and Mehdi-Souzani C 2016 Improved curvature-based registration methods for high-precision dimensional metrology *Precis. Eng.* **46** 232–42

Ren M J, Cheung C F and Kong L B 2012 A task specific uncertainty analysis method for least-squares-based form characterization of ultra-precision freeform surfaces *Meas. Sci. Technol.* **23** 054005

Rouhani M, Sappa A D and Boyer E 2014 Implicit B-spline surface reconstruction *IEEE Trans. Image Process.* **24** 22–32

Savio E and De Chiffre L 2002 An artefact for traceable freeform measurements on coordinate measuring machines *Precis. Eng.* **26** 58–68

Savio E, De Chiffre L and Schmitt R 2007 Metrology of freeform shaped parts *Ann. CIRP* **56** 810–35

Saux E and Daniel M 2003 An improved Hoschek intrinsic parametrization *Comput. Aided Geom. Des.* **20** 513–21

Sharp G C, Lee S W and Wehe D K 2002 ICP registration using invariant features *IEEE Trans. Pattern Anal. Mach. Intell.* **24** 90–102

Sui X L, Guan Y, Zhang S G and Chen B 2016 Freeform surface profile error evaluation based on gray code genetic algorithm and segmentation of spherical approximation method *2015 Sixth Int. Conf. on Intelligent Systems Design and Engineering Applications (ISDEA) (Guiyang, 2015)* pp 410–3

Tan G, Zhang L, Liu S and Zhang W 2015 A fast and differentiated localization method for complex surfaces inspection *Int. J. Precis. Eng. Man.* **16** 2631–9

Tang J, Fang M E and Wang G 2018 Subdivision of uniform ωB-spline curves and two proofs of its Ck-2-continuity *Comput. Model Eng. Sci.* **115** 263–80

Tjahjowidodo T, Dung V T and Han M L 2015 A fast non-uniform knots placement method for B-spline fitting 2015 *IEEE AIM (Busan, South Korea, July)* pp 1490–5

Wang W, Pottmann H and Liu Y 2006 Fitting B-spline curves to point clouds by curvature-based squared distance minimization *ACM Trans. Graph.* **25** 214–38

Weiss V, Andor L, Renner G and Várady T 2002 Advanced surface fitting techniques *Comput. Aided Geom. Des.* **19** 19–42

Whitehouse D J 2011 *Handbook of Surface and Nanometrology* (Boca Raton, FL: CRC Press)

Yamany S M and Farag A A 2002 Surface signatures an orientation independent free-form surface representation scheme for the purpose of objects registration and matching *IEEE Trans. Pattern Anal. Mach. Intell.* **24** 1105–20

Yan J, Zhang Z, Kuriyagawa T and Gonda H 2010 Fabricating micro-structured surface by using single-crystalline diamond endmill *Int. J. Adv. Manuf. Technol.* **51** 957–64

Yang H, Wang W and Sun J 2004 Control point adjustment for B-spline curve approximation *Comput. Aided Des.* **36** 639–52

Yau H T and Menq C H 1996 A unified least-squares approach to the evaluation of geometric errors using discrete measurement data *Int. J. Mach. Tools Manuf.* **36** 1269–90

Zhang D and Lu G 2002 Generic Fourier descriptor for shape-based image retrieval *Proc. IEEE Int. Conf. Multimedia Expo. (Lausanne, Switzerland, August)* pp 425–8

Zhang D, Maropoulos P G and Hill M 2006 Random uncertainty propagation in estimates of sphere parameters from coordinate measurements *Int. J. Mach. Tools Manuf.* **46** 1362–8

Zhang L, Wang H, Ge X and Tan J 2015 Dual generalized B-spline functions and their applications in several approximation problems *J. Adv. Mech. Des. Syst. Manuf.* **9** 0053

Zhang X 2009 Free-form surface fitting for precision coordinate metrology *Doctoral dissertation* University of Huddersfield

Zhang Y, Cheng H N, Wu R and Liang R 2017 Data processing for point-based *in situ* metrology of freeform optical surface *Opt. Express* **25** 13414–24

Zhu K P, Wong Y S, Lu W F and Fuh J Y 2009 A diffusion wavelet approach for 3-D model matching *Comput. Aided Des.* **41** 28–36

Chapter 8

Performance verification for optical co-ordinate metrology

Adam Thompson and Nicholas Southon

In this chapter, a summary is provided of documents relevant to the performance verification of optical co-ordinate measurement systems, primarily ISO 10360 part 13 and VDI/VDE 2634. The concept of performance verification is introduced and its usage is described. Appropriate material measures for performance verification of optical co-ordinate measurement systems are described, the different types of maximum permissible error are discussed and procedures for verifying systems are explained. Finally, the validity of performance verification exercises is discussed, explaining where performance verification is and is not applicable.

8.1 Introduction

The *Vocabulary in Metrology* (VIM) describes *verification* as the 'provision of objective evidence that a given item fulfils specified requirements' (BIPM 2012). Consequently, as discussed in chapter 1, *performance verification* is used to confirm that a measurement system meets its specification. In co-ordinate measurement, performance verification is generally performed in reference to the system specific part of the ISO 10360 series of standards (ISO 2000). In the case of optical 3D co-ordinate measurement systems (CMSs), the appropriate part is the as-yet unpublished ISO 10360 part 13 (ISO 2020). Since the ISO 10360 part 13 standard is currently at draft international standard stage, the VDI/VDE 2634 parts 2 (VDI/VDE 2012) and 3 (VDI/VDE 2008) guidelines are commonly used for performance verification for optical 3D CMSs. VDI/VDE 2634 part 2 covers area scanning optical 3D CMSs which measure from a single measurement pose ('single view'). VDI/VDE 2634 part 3 is similar to VDI/VDE 2634 part 2, but covers measuring from multiple measurement poses ('multi-view'). When published, ISO 10360 part 13 will cover both single-view and multi-view measurements.

doi:10.1088/978-0-7503-2524-0ch8

In this chapter, methods for performance verification will be described in reference to both VDI/VDE 2634 and ISO 10360 part 13. These methods can be used to verify the performance of any optical CMS. Because of their relative similarity (only being differentiated by single- against multi-view measurement set-ups), for the purpose of this chapter, VDI/VDE 2634 parts 2 and 3 will be generically referred to, hereafter, as VDI/VDE 2634.

8.2 Material measures

A series of calibrated material measures are required to verify the performance of an optical CMS. Appropriate material measures used to perform performance verification are specified in the relevant guidance document/standard, be that VDI/VDE 2634 or ISO 10360 part 13. The actual material measures used to complete any one performance verification exercise are then chosen to meet the requirements of the relevant guidance document/standard. It is usually possible to use a single set of material measures to satisfy the requirements of both performance verification methods. The material measures used for both VDI/VDE 2634 or ISO 10360 part 13 are presented in this section.

There are a number of factors to consider when designing or choosing material measures. Particularly, the size of the deviation from nominal of any material measure feature must be negligibly small compared to the capabilities of the instrument being verified (i.e. the material measure can be considered nearly 'perfect', see sections 8.2.1–8.2.4) and the material of the measured features must be opaque, to prevent sub-surface reflection of light. All measured distances used in the performance verification process must be traceably measured and presented with an expanded uncertainty. Practical mounting of the material measures should be designed so that the fixtures are not in contact with measured features, and so that minimal stress is placed on the material measure (Carmignato *et al* 2020).

In practice, it is likely that instrument users will combine the material measures required into a smaller number of measurement artefacts that fulfil the requirements of the relevant document while minimising the number of physical objects required to perform a performance verification. Material measures could also take alternative forms to those presented here (for example, ball plates instead of ball bars/ball beams). In this chapter, however, examples of the required material measures are presented in their most basic form, without combining the measures into a set of measurement artefacts or providing additional measurement features or options.

8.2.1 Material measure 1

Material measure 1 is used to test the probing characteristics (see section 8.5.3) of the optical CMS and takes the form of a calibrated sphere. The sphere should have negligibly small sphericity (i.e. the deviation from a mathematical sphere) compared to the expected probing characteristics of the instrument being tested. The value at which this deviation becomes negligible will depend on each individual optical CMS, but as a guideline, the deviation of the real sphere from a mathematical sphere should be no larger than a few micrometres (based on the state of the art in optical co-ordinate measurement at the time of publication of this book). The characteristics of the material used to create the spheres will depend on the type of instrument being tested. To use fringe projection as an example, the spheres must also be of a 'matte' finish, as fringe projection systems are generally not capable of measuring polished surfaces (see chapter 5 and Zhang 2018). Other optical set-ups, however, will demand different material characteristics. Material measure 1 is shown in figure 8.1 and the dimensional constraints for the nominal length used are presented in section 8.5.3.

8.2.2 Material measure 2

Material measure 2 is used to test the distortion characteristics (see section 8.5.4) of the optical co-ordinate measurement system. Material measure 2 takes the form of a ball bar: two spheres attached to either end of a rod. As with material measure 1, the spheres attached to each end of the rod should have negligibly small sphericity (i.e. the deviation from a mathematical sphere) compared to the expected probing characteristics of the instrument being tested. Again, the value at which this deviation becomes negligible will depend on each individual optical CMS, but as a guideline, the deviation of the real sphere from a mathematical sphere should be no larger than a few micrometres (based on the state of the art in optical co-ordinate measurement at the time of publication of this book). As with the spheres used in material measure 1, material characteristics for the spheres depend on the type of optical CMS being tested, but the rod should be made from a material stiff enough to prevent measureable deformation during the performance verification process. Material measure 2 is shown in figure 8.2 and the dimensional constraints for the nominal lengths used are presented in section 8.5.4.

Figure 8.1. Material measure 1: a calibrated sphere.

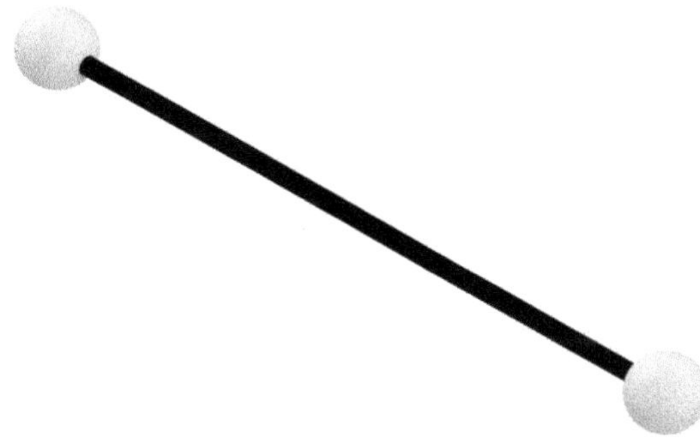

Figure 8.2. Material measure 2: a rod with spheres affixed to either end.

Figure 8.3. Material measure 3: a flat plate.

8.2.3 Material measure 3

Material measure 3 is used to test the distortion characteristics (see section 8.5.4) of the optical CMS, and comprises a flat plane. The surface of the plane should have a negligibly small flatness (i.e. the deviation from an ideal flat plane) compared to the expected probing characteristics of the instrument being tested. While the value at which the deviation becomes negligible differs for each individual optical CMS, as a rule of thumb, the peak-to-valley deviation should be no larger than a few micrometres. As with the spheres used in material measures 1 and 2, material characteristics for the spheres depend on the type of optical CMS being tested, but the rod should be made from a material stiff enough to prevent measureable deformation during the performance verification process. Material measure 3 is shown in figure 8.3 and the dimensional constraints for the nominal lengths used are presented in section 8.5.4.

8.2.4 Material measure 4

Material measure 4 is used to test the length measurement errors of the optical CMS (see section 8.5.5). This material measure comprises a multi-ball beam with at least five independent test lengths, the longest length approximating the diagonal of the *concatenated measurement volume*, where the concatenated volume is the 'volume in which the sensor can determine measurement data including movement of the sensor and registration of measurement data obtained in different positions and orientations of the sensor relative to the artefact fulfilling the specifications provided by the manufacturer' (ISO 2020). For this material measure, the simplest artefact that can be used is a ball beam with six spheres; the spheres used conforming to the same requirements discussed in reference to material measures 1 and 2. As with material measure 2, the beam should be made from a material stiff enough to prevent measurable deformation during the performance verification process. In this case, the user should note that as this material measure is likely to be relatively large (concatenated volumes are often a couple of metres along the diagonal), the stiffness of the rod should be particularly considered. Material measure 4 is shown in figure 8.4 and the dimensional constraints for the nominal lengths used for the material measure are presented in section 8.5.5.

8.3 The different types of performance verification test

According to ISO 10360 part 13, performance verification is required in three scenarios. These scenarios are 'acceptance tests', 'reverification tests' and 'interim checks'. In VDI/VDE 2634, reverification tests and interim checks are both referred to as 'reverification checks'. These scenarios are described in this section.

To pass performance verification tests, measurements must generally conform to a set of *maximum permissible errors* (MPEs) for the specified measurement procedures. A MPE is defined in the VIM as an 'extreme value of measurement error, with respect to a known reference quantity value, permitted by specifications or regulations for a given measurement, measuring instrument, or measuring system' (see chapter 1 and BIPM 2012).

An acceptance test is performed upon purchasing, maintaining, repairing, renovating or upgrading an instrument, and it is commonly a contractual obligation between the instrument user and the instrument manufacturer. The acceptance test covers all of the individual tests described in this chapter, as well as any user-specific performance verification tests outlined in relevant agreements between the user and the manufacturer. During acceptance tests and any other measurements, the measurement system will be operated according to the manufacturer's specifications

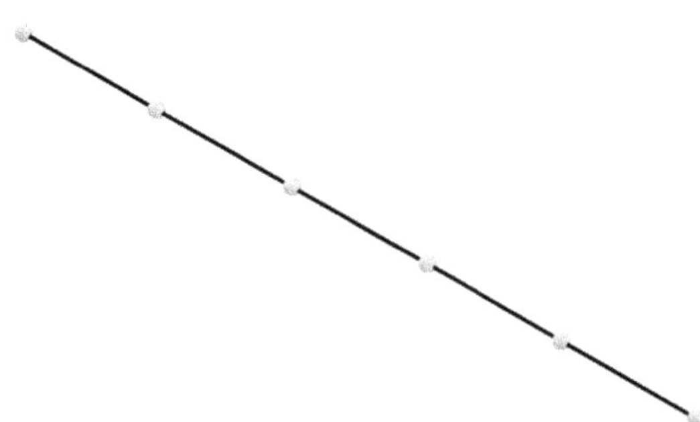

Figure 8.4. Material measure 4: a multi-ball artefact containing at least five independent test lengths.

and procedures, in line with the methods described in ISO 10360 part 13, VDI/VDE 2634 and this chapter.

A reverification test is part of an organisation's internal quality assurance policy and includes the same methodology as the acceptance test. The reverification test confirms whether the measurement system is operating within its specified parameters. Reverification tests are always performed according to the user's specifications (see section 8.4.1) and the manufacturer's procedures.

An interim check is part of an organisation's internal quality assurance system and is recommended as a periodic (for example, daily, weekly) indication that the measurement system is operating within its specified parameters; with measurements conforming to the defined set of MPEs. Compared to the acceptance and reverification tests, an interim check generally features a lower number of measurements, to reduce the time necessary to perform the check. These measurements are often a subset of those required by the acceptance and reverification tests. A failed interim check is generally cause for a full reverification test.

The frequency of interim checks and the checks themselves will vary depending on system and user, but checks should be performed as often as is feasible, given the user's time constraints. For example, each week could begin with an interim check lasting around 1 h, while each measurement could contain an *in situ* check that can be completed in a few minutes. An example weekly interim check using a subset of the reverification procedure may involve performing the measurement volume distortion error test (see section 8.5.4.1), while an example *in situ* check may involve measurement of a calibrated sphere within the measurement volume, during each measurement.

8.4 Specification of errors

When performing a full performance verification exercise for any particular instrument, there are often a large number of MPEs that must be tested (twenty-one combinations of five types of errors and both data selection and measurement style options for the ISO case (ISO 2020), as well as the four VDI/VDE errors (VDI/VDE 2012, VDI/VDE 2008)). Some MPEs are optional, but both the VDI/VDE and ISO documents recommend that for good practice, all MPEs are evaluated. MPE naming conventions differ between the two documents, but often refer to the same or similar actual quantities. The selection of MPEs are named in ISO 10360 part 13 to sit within the whole of the ISO 10360 family of standards, while the VDI/VDE naming convention offers a simpler, if less transferrable system of nomenclature. The MPEs presented in both ISO 10360 part 13 and VDI/VDE 2634 are summarised in table 8.1. In table 8.1, the *i* modifiers '95%', 'All' and '1×25', refer to the three different data use conditions using, respectively, 95% of the acquired data by excluding the outlying 5% of points, all of the acquired data or twenty five representative points (see section 8.5.1). The *j* modifiers, 'SMV' and 'CMV', refer, respectively, to sensor measurement volume and concatenated measurement volume, while the *j* modifiers 'SV' and 'MV' refer, respectively, to 'single-view' (i.e. a single measurement shot) and MV 'multi-view' (i.e. data fused from multiple single

Table 8.1. MPEs specified in ISO 10360 part 13 and VDI/VDE 2634.

Name of error	ISO symbol(s)	VDI/VDE symbol	'*i*' modifiers possible	'*j*' modifiers possible
Probing form error	$P_{\text{Form.Sph.}i:j:\text{O3D}}$	*PF*	95%, 97% All, 1 × 25	SMV.SV, SMV.MV
Probing size error	$P_{\text{Size.Sph.}i:j:\text{O3D}}$	*PS*	95%, 97% All, 1 × 25	SMV.SV, SMV.MV
Distortion error	$D_{\text{CC.}j:\text{O3D}}$	*SD*		SMV.SV, SMV. MV, 97%:CMV
Flat form distortion error	$D_{\text{Form.Pla.}i.j:\text{O3D}}$		95%, All	SMV.SV, SMV.MV
Length measurement error	$E_{i:j:\text{O3D}}$	*E*	MVol, Bi, Uni, Bi(VDI/VDE)	CMV.MV

Table 8.2. Calibrated dimensions required for the measurements.

Name of dimension	Symbol	Material measure
Centre-to-centre distance of ball bar	$L_{\text{CC.cal}}$	Two ball bar
Diameter of ball bar balls	d_{cal}	Two ball bar
Five different centre-to-centre distances on ball bar	$L_{\text{CC.cal}}$	Six ball beam

views). Throughout the remainder of this chapter, the more complex (but also more complete and more transferable) ISO nomenclature is used.

Each MPE in table 8.1 relates to a calibrated dimension. In table 8.2, the calibrated dimensions required when calculating the error values for each measurement are summarised.

8.4.1 General guidelines for performance verification

When carrying out any performance verification task using optical CMSs, there are several general considerations that should be applied in order to conform to the relevant guidance document/standard. These considerations are noted in this chapter where relevant, but are explicitly as follows:

- Additional or omitted measurement instrument set-ups (for example, hardware and/or software settings) can influence measurements (for example, using more or fewer set-ups, altering lighting/processing settings). Instrument manufacturers require that users have enough knowledge about these influences to minimise their negative impacts on measurements.
- Optical co-ordinate measurements generally rely on light reflect from the measured surface. Therefore, surface characteristics such as colour, glossiness and texture can limit measurements.

- When verifying the performance of any one optical CMS, the material and surface texture of the material measures are stated by the system manufacturer.
- Manual and/or automatic repositioning of the measured object or the measurement system can be used for any measurements. The position and/ or orientation can be used as additional information for the registration of multiple-view measurements.
- Acceptance and reverification tests are to be conducted using the same operating procedures and within the same environment as day-to-day measurements made using the system.
- The use of software functions for the automatic removal of measured points that do not belong to the measured surface is permitted by the relevant guidance document/standard. If the selection of points for evaluation is manual, and part of the normal operating procedure, these operations are to be detailed in the rated operating conditions and implemented for the test. Automated or manual data selection should not be used to eliminate noise or outliers from the data.
- All data not excluded by the automatic and manual methods described above should be used for measurements. The derivation of test errors for probing form error or probing size error for both single-view and multi-view measurements (see section 8.5.3), can encompass either 95% or 100% of the measured data. Errors are to be recorded for both the 95% and 100% cases. Flatness errors must be calculated for both 95% and 100% of the data.
- The relative length scale used to define the majority of the dimensional requirements of the material measures is the longest spatial diagonal of the measurement volume.

8.5 Performance verification procedures

In this section, procedures for conducting performance verification in line with both ISO 10360 part 13 and VDI/VDE 2634 are covered. Because of the relative similarity between the two procedures, they are presented as a single summative procedure, with discrepancies between them noted where relevant.

The overall performance verification test is divided into three primary areas, describing methods for verification of probing characteristics, distortion characteristics and length measurement errors, respectively. Prior to describing these aspects of the test in detail, however, a method for determining ‘representative points’ for sphere measurement is provided (see section 8.5.1).

In earlier drafts of ISO 10360 part 13, determination of representative points was required to compare part 13 of the 10360 series with the other parts of ISO 10360, because of the intrinsic nature of the contact measurement systems described by the earlier parts of ISO 10360. Particularly, the low numbers and density of acquired points that older contact-based CMSs is noted, so the number of points acquired by optical CMSs (which typically acquire much larger numbers of points at much higher density), is reduced to allow comparison to contact systems.

At the time of publication of this book, there is an ongoing debate about whether the representative points process is required for performance verification of optical CMS. In the most recent version of ISO 10360 part 13, the representative points requirement has been removed. However, to account for the eventuality where this section is returned to the document at some later date, the method is summarised here.

VDI/VDE 2634 does not allow for representative points to be used for measurements.

8.5.1 Measurements with representative points

A method for generating representative points on a surface is presented, for use when calculating probing characteristics (as per ISO 10360 part 8 ISO (2013)). This method is intended to enable comparability of these tests with similar tests described in other parts of ISO 10360, and involves reducing a point cloud to a set of 25 representative points.

The twenty five points must be approximately evenly distributed over at least a hemisphere. The ISO-recommended probing points are defined in five groups and illustrated in figure 8.5. The first group is a single point on the pole of the sphere (the pole location is determined by the probing direction of the measurement). Each of the four remaining groups of points lies on equally spaced contours of constant polar angle. In the case of a hemisphere, this means each group is 22.5 further below the pole than the previous group. The second group contains four points on the polar contour. The third group contains eight points, the fourth contains four points and the fifth group contains eight points, respectively.

Representative points are calculated from patches of measured data no larger than 2% of the measured area of the sphere. These patches are generated from a series of locations on the sphere, defined in ISO 10360 part 5 (ISO 2010). There are

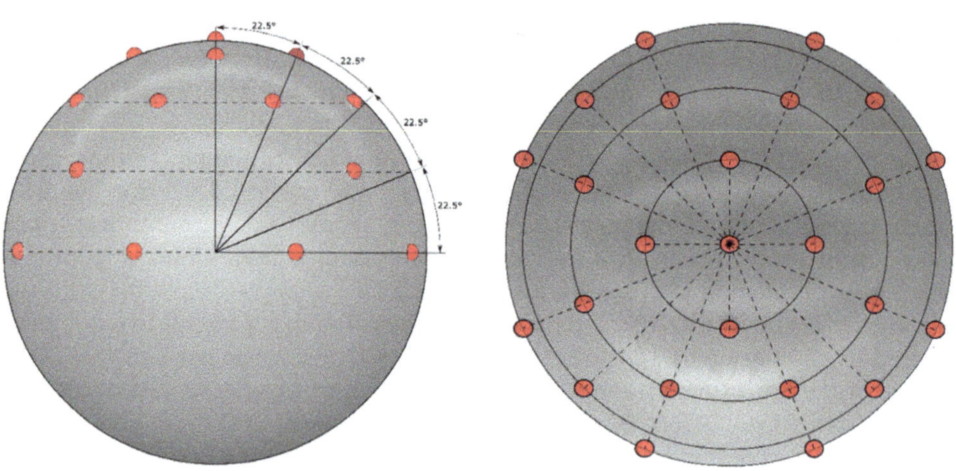

Figure 8.5. Side view and top view diagrams detailing the selection of locations to be used to calculate representative points.

multiple methods by which representative points can be derived from measurement data; the method described here was first described by Borges de Olivera *et al* (2015). This particular method is based on the approach presented in ISO 10360 part 8 (ISO 2013), with further clarification provided to facilitate practical application of the method (see figure 8.6).

- Following the acquisition of point cloud data, data points that are from surfaces that are not part of the sphere are removed. Next, an unconstrained least-squares sphere is fitted to all data points corresponding to the measured sphere.
- Representative point positions are chosen on the fitted sphere.
- For each patch centre, a cone is constructed with its apex at the centre of the sphere fitted to the measured data.
- All measured points that intersect with a patch's cone are classified as being part of that patch.
- For each patch, a least-squares sphere is fitted using the reduced point cloud.
- For each patch, a line is drawn that passes through both the centre of the sphere defined in the previous step and the centre of mass of the patch's measured data.
- The point of intersection of a patch's fitted sphere surface with the aforementioned line becomes the representative point for that patch.

8.5.2 General preparation for the performance verification test

Prior to performing any measurements, ISO 10360 part 13 and VDI/VDE 2634 specify that a preparation routine should be performed to ensure that the specified environmental and operating conditions are adhered to. For acceptance tests, the environmental conditions are specified by the manufacturer and for reverification tests they are specified by the user, although user specified conditions cannot fall outside of the manufacturer specified conditions. Environmental conditions are specified by limits on the permissible conditions in which measurements can be made. Examples of specified conditions include temperature, air humidity, vibration and lighting.

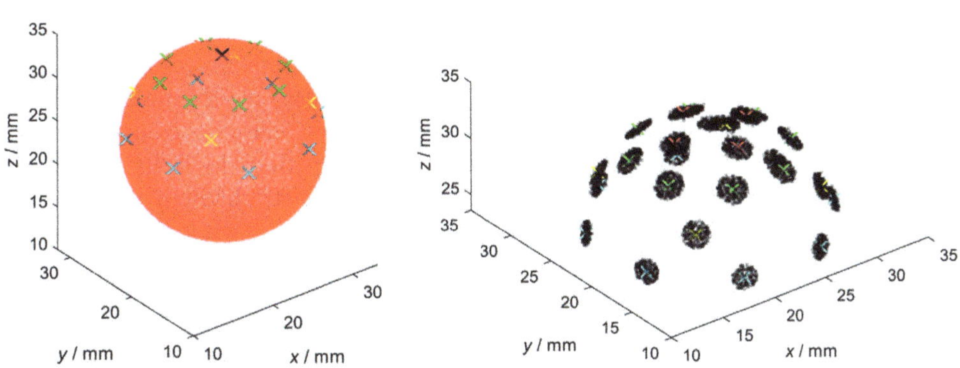

Figure 8.6. A point cloud of a sphere being subsampled into 25 representative points.

In order for the instrument to be considered as operating within its expected parameters, the manufacturer's operating conditions must be adhered to for all measurements made. These conditions are generally found in the manufacturer's operating manual and specific areas to adhere to may include:

- machine start-up/warm-up cycles;
- probing system qualification;
- thermal stability of the measurement system before the system qualification;
- location, type and number of thermal sensors;
- software filters;
- surface characteristics of material measures which may include colour, glossiness or texture; and
- default procedures and settings for data registration and data fusion.

8.5.3 Probing characteristics

The probing characteristics of an optical CMS can be acquired by measurement of a single sphere (material measure 1). There are two types of error that make up the probing characteristics of an optical CMS. The 'probing size error' of a sphere is the difference in radius between a sphere fitted to the measured data and the calibrated radius of the measured sphere, while the 'probing form error' of a sphere is the thickness of a spherical shell which encompasses the measured data. Assuming the sphere used has negligible distortion from a mathematical sphere, compared to the expected probing characteristics of the instrument being tested (see section 8.2.1), measurements of the sphere allow the user to determine these characteristics.

8.5.3.1 Dimensional requirements of the material measure

The material measure for this test consists of a calibrated test sphere. The test sphere diameter D_p must satisfy the following conditions:

$$0.02\, L_0 \leqslant D_p \leqslant 0.2\, L_0,$$

where L_0 is the longest spatial diagonal of the sensor volume.

8.5.3.2 Procedure

If following the method prescribed by ISO 10360 part 13, to ensure a uniform sampling of the sensor measurement volume, the measurement volume is divided into eight sub-volumes (or 'voxels') of similar size. The test for probing errors is then performed for each voxel, with at least three repeat measurements made in each position. The measurable part of the sphere must be positioned wholly in the voxel in which it is meant to be located. Using data obtained through a single-view measurement, $P_{X.\mathrm{Sph}.i:\mathrm{SMV.SV:O3D}}$ errors can be evaluated, while data obtained through a multiple-view measurement can be used to evaluate $P_{X.\mathrm{Sph}.i:\mathrm{SMV.MV:O3D}}$.

If following the method prescribed by VDI/VDE 2634, measurements should be made of a sphere placed in at least three evenly distributed positions across the measurement volume. For each sphere position, data should be captured from at least five measurement observation positions (i.e. the position of the optical CMS with respect to the sphere). VDI/VDE 2634 does not require any repeats of these measurements. The measurement observation positions are suggested by VDI/VDE 2634 to be five of the six cardinal positions around the measured sphere, as shown in figure 8.7.

8.5.3.3 Calculation of errors

Probing form error

For the calculation of all ISO 10360 part 13 probing errors, both 95% and 100% of the data should be used. When using 95% of the data, the user should exclude the 5% of data furthest from the mean value, to minimise the error. The same data is used to calculate both errors. The following probing form errors are defined:

- $P_{\text{Form.Sph.1×25:SMV.SV:O3D}}$, probing form distortion error, 25 representative points, single view;
- $P_{\text{Form.Sph.95\%:SMV.SV:O3D}}$, probing form distortion error, 95% data, single view;
- $P_{\text{Form.Sph.All:SMV.SV:O3D}}$, probing form distortion error, 100% data, single view;
- $P_{\text{Form.Sph.1×25:SMV.MV:O3D}}$, probing form distortion error, 25 representative points, multi-view;
- $P_{\text{Form.Sph.95\%:SMV.MV:O3D}}$, probing form distortion error, 95% data, multi-view; and
- $P_{\text{Form.Sph.All:SMV.MV:O3D}}$, probing form distortion error, 100% data, multi-view.

The probing form distortion error is the width of a spherical shell that contains either 95% or 100% of the measurement data, depending on the error being calculated. For example, when calculating $P_{\text{Form.Sph.95\%:SMV.SV:O3D}}$, a sphere is fitted to all the measurement data. The centre of this fitted sphere is used to construct a spherical shell that contains the 95% of data that is closest to the average radius of

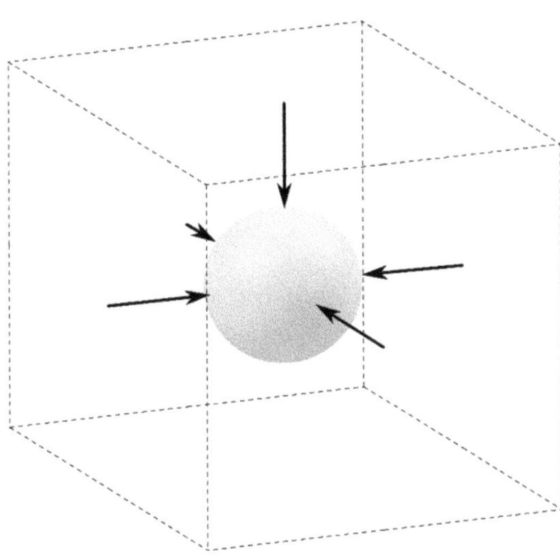

Figure 8.7. The five probing directions suggested by VDI/VDE 2634.

the fitted sphere. The shell, which contains the chosen 95% of the data, is described by two radii: a minimum radius r_{min} and a maximum radius r_{max}. The difference between the two radii of the shell is $P_{\text{Form.Sph.95\%:SMV.SV:O3D}}$.

For calculation of errors per VDI/VDE 2634, 3% of the data can be excluded when fitting a sphere to the data. The calculation of form error remains the same. In VDI/VDE 2634, this error is notated as *PF*, but has been amended for consistency within this chapter to conform to the ISO 10360 part 13 convention. As such, the error is notated here as follows:

- $P_{\text{Form.Sph.97\%:SMV.SV:O3D}}$, probing form distortion error, 97% data, single view.

Probing size error

The probing size error is the difference between the measured diameter of the sphere d_{meas} and the calibrated diameter of the sphere d_{cal}, i.e.

$$P_{\text{Size.Sph.}i:j\text{:O3D}} = d_{\text{meas}} - d_{\text{cal}}.$$

As in the probing form error case, when using 95% of the data, the user should exclude the 5% of data furthest from the mean value, to minimise the error. The following probing size errors are defined:

- $P_{\text{Size.Sph.95\%:SMV.SV:O3D}}$, probing size error, 95% data, single view;
- $P_{\text{Size.Sph.1}\times\text{25:SMV.SV:O3D}}$, probing size error, 25 representative points, single view;
- $P_{\text{Size.Sph.All:SMV.SV:O3D}}$, probing size error, 100% data, single view;
- $P_{\text{Size.Sph.1}\times\text{25:SMV.MV:O3D}}$, probing size error, 25 representative points, multi-view;
- $P_{\text{Size.Sph.95\%:SMV.MV:O3D}}$, probing size error, 95% data, multi-view; and
- $P_{\text{Size.Sph.All:SMV.MV:O3D}}$, probing size error, 100% data, multi-view.

As in the probing form error case, for calculation of errors per VDI/VDE 2634, 3% of the data can be excluded when fitting a sphere to the data. The calculation of size error remains the same. In VDI/VDE 2634, this error is notated as *PS*, but has been amended for consistency within this chapter to conform to the ISO 10360 part 13 convention. As such, the error is notated here as follows:

- $P_{\text{Size.Sph.97\%:SMV.SV:O3D}}$, probing size distortion error, 97% data, single view.

8.5.4 Distortion characteristics

The distortion characteristics of the optical CMS can be acquired by measuring a ball bar and a flat plane (i.e. material measures 2 and 3). Measurement of these material measures allows a user to determine the general 'distortion error' for the measurement volume, as well as the 'flat form distortion'. For ISO 10360 part 13, the distortion error is the error of indication when measuring a sphere centre-to-centre distance within the sensor measurement volume, whilst the VDI/VDE document requires a concatenated measurement volume to be used. The flat form distortion error is the smallest distance between two parallel planes that encompasses all measured data obtained for a test flat.

8.5.4.1 Measurement volume distortion error

The assessment of the measurement volume distortion error uses evenly distributed length measurements to reveal distortion in the measurement volume.

Dimensional requirements of material measures

The material measure for measurements made per ISO 10360 part 13 consists of two spheres separated by a known distance (i.e. material measure 2). The centre-to-centre distance of the spheres is L_p, the nominal diameter of the spheres is D_p, and the longest spatial diagonal of the sensor volume is L_0. There is an upper limit on L_p from the practical consideration of the material measure being contained within a single (i.e. not concatenated) measurement volume for all measurement positions (see figure 8.8 and table 8.3). The value of the upper limit is dependent on the

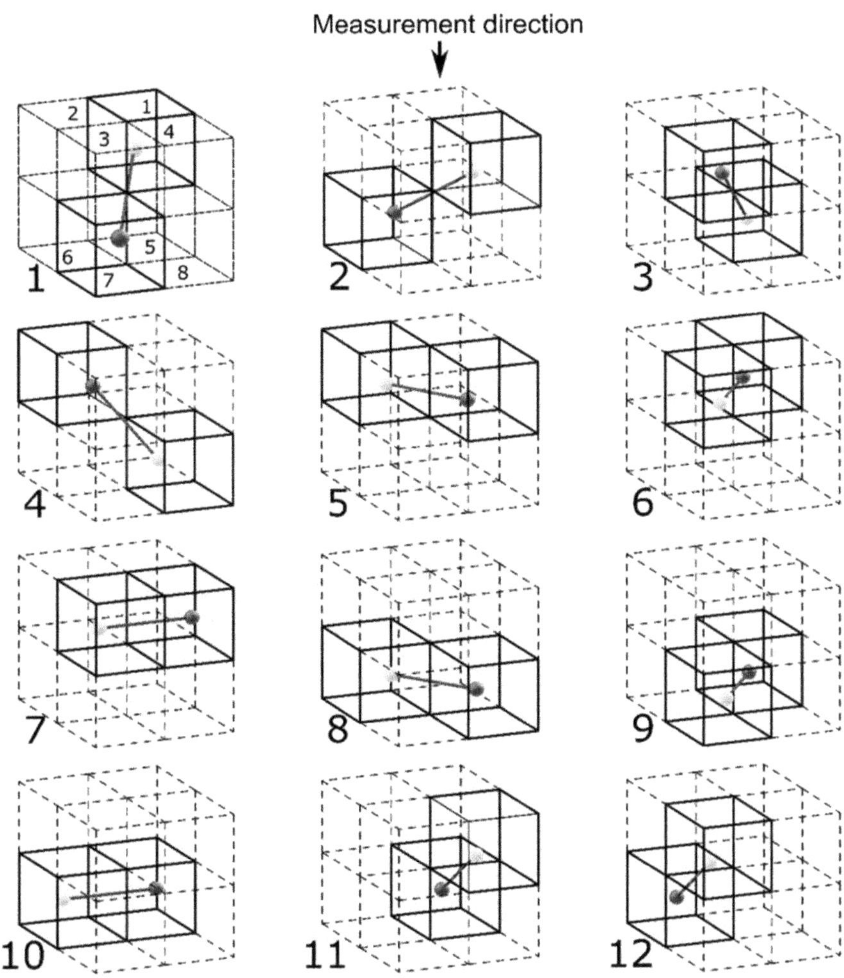

Figure 8.8. A diagram of the twelve poses to be measured per ISO 10360 part 13. Voxels are numbered for pose 1 but apply throughout the figure and this chapter.

Table 8.3. Ball bar poses required by ISO 10360 part 13.

Pose	Sphere 1 in voxel	Sphere 2 in voxel
ISO 1	1	7
ISO 2	4	6
ISO 3	5	3
ISO 4	8	2
ISO 5	2	4
ISO 6	3	1
ISO 7	3	4
ISO 8	6	8
ISO 9	7	5
ISO 10	6	5
ISO 11	4	5
ISO 12	3	6

geometry of the measurement volume and the sub-division of the area (see section 8.5.3.2). The lower limit on L_p and the constraints on D_p are as follows:

$$L_p \geqslant 0.3\, L_0; \text{ and}$$
$$0.02\, L_0 \leqslant D_p \leqslant 0.2\, L_0.$$

The constraints on L_p can be difficult to satisfy if the measurement volume is far from an ideal cube. If the ratio of the measurement volume's longest side length to the shortest side length is larger than three, per ISO 10360 part 13 the dimensions of the material measure can be adopted by mutual agreement between the user and the instrument manufacturer.

For measurements made per VDI/VDE 2634, more than one length must be used, because of the constraints placed on the test length by the specification. The constraints on D_p are the same as in the ISO 10360 part 13, but in the VDI/VDE document, the constraint for L_p is different, and in fact has two separate cases. When measuring VDI/VDE positions 1–3 (see figure 8.9 and table 8.4) the constraint is as follows:

$$L_{p1} \geqslant \frac{2}{3} L_{0:\text{CMV}},$$

where $L_{0:\text{CMV}}$ is the longest spatial diagonal of the concatenated measurement volume. When measuring VDI/VDE position 7 (see figure 8.9 and table 8.4) the constraint is as follows:

$$L_{p2} \geqslant \frac{2}{3} L_{\text{min:CMV}},$$

where $L_{\text{min:CMV}}$ is the length of the shortest side of the concatenated measurement volume. When measuring VDI/VDE positions 4–6 (see figure 8.9 and table 8.4), the

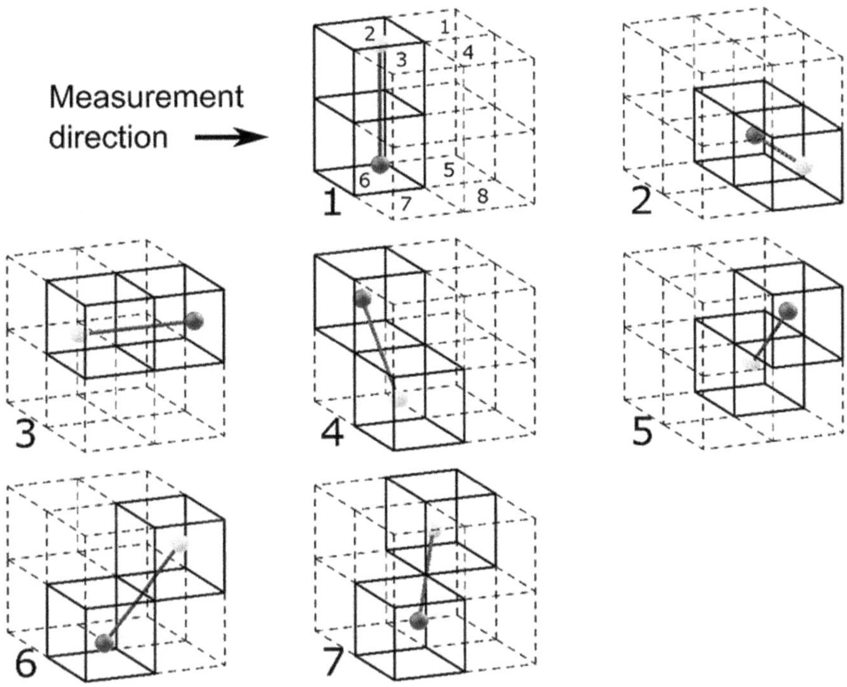

Figure 8.9. A diagram of the seven poses to be measured per VDI/VDE 2634. Voxels are numbered in pose 1 but apply throughout the figure and this chapter (as per figure 8.8).

Table 8.4. Ball bar poses required by VDI/VDE 2634.

Pose	Sphere 1 in voxel	Sphere 2 in voxel
VDI/VDE 1	2	6
VDI/VDE 2	8	5
VDI/VDE 3	3	4
VDI/VDE 4	7	2
VDI/VDE 5	5	4
VDI/VDE 6	4	7
VDI/VDE 7	1	7

longer distance L_{p1} should be used if the whole feature fits within the measurement volume. The example provided in this chapter for material measure 4 (see section 8.2.4) has a long sphere centre-to-centre distance that is suitable for L_{p1}, as well as a short sphere centre-to-centre distance that could be suitable for L_{p2}. This is not the case when a small value of $L_{\text{min:CMV}}$ means the required length for L_{p2} is shorter than material measure 4 can provide. If

$$L_{\text{min:CMV}} \lesssim 25\% \, L_{0\text{:CMV}},$$

then an alternative material measure will likely be needed.

Procedure
To ensure uniform sampling of the sensor measurement volume, the measurement volume is again divided into eight voxels of similar size. For measurements made per ISO 10360 part 13, the measurable parts of each of the spherical features of the material measure must be positioned wholly in different voxels. There are twenty eight unique positions possible that fulfil this criterion, but there are 12 orientations that must be measured to fulfil the requirements of the standard (see table 8.3 and figure 8.8). In addition, each voxel must have at least one measurement where the measurable portion of a sphere is no further than 10% of L_0 from the boundary. Depending on the circumstances of the measurements, this requirement may be met naturally, but the user is allowed by ISO 10360 part 13 to intervene to ensure this requirement is met where necessary.

Single-view measurements
In ISO 10360 part 13, there are 12 compulsory measurement positions and each measured orientation of the material measure requires three repeat measurements, giving a total of 36 measurements.

Multi-view measurements
Per ISO 10360 part 13, the 12 compulsory measurement positions are then registered and fused into one point cloud in accordance with the manufacturer's guidelines. In line with the standard, this operation should be performed separately for each of the required three repeat measurements, giving three fused multi-view point clouds comprised of data from 12 views.

Concatenated view measurements
VDI/VDE 2634 states a preference for measurements to be made in a concatenated measurement volume when testing distortion error. As per the guidance, the longest diagonal of the concatenated measurement volume should be at least twice as long as the longest length of the single measurement volume. The seven specified material measure locations and orientations are shown in figure 8.9 and specified in table 8.2 as VDI/VDE 1–7. No repeat measurements are required to meet the VDI/VDE requirements.

Calculation of distortion errors
In ISO 10360 part 13, a distortion error $D_{\mathrm{CC:SMV.SV:O3D}}$ is obtained for each of the 36 single-view measurements. The multi-view distortion error $D_{\mathrm{CC:SMV.MV:O3D}}$ is obtained for each of the three fused measurements. For every measurement, a sphere is fitted to all the data of both spheres on the ball bar. If different fitting options exist in the manufacturer's software, ISO 10360 part 13 specifies that the option that minimises the maximum deviation from the ideal feature should be chosen (commonly referred to as Chebychev fitting (Drieschner 1993)). The distance between the centres of the fitted spheres is $L_{\mathrm{CC.meas}}$. Both errors are calculated from the difference between the measured and calibrated values of the spheres' centre-to-centre distances ($L_{\mathrm{CC.cal}}$):

$$D_{\text{CC}.j:\text{O3D}} = L_{\text{CC.meas}} - L_{\text{CC.cal}}; \text{ and}$$
$$D_{\text{CC}.j:\text{O3D}} = L_{\text{CC.meas}} - L_{\text{CC.cal}}.$$

The following probing size errors are defined:

- $D_{\text{CC.SMV.SV:O3D}}$, single-view distortion error; and
- $D_{\text{CC.SMV.MV:O3D}}$, multi-view distortion error.

For calculation of distortion error per VDI/VDE 2634, 3% of the data can be excluded when fitting spheres to the data. The calculation of distortion error remains the same. In VDI/VDE 2634, this error is referred to as sphere spacing error and is notated as *SD*, but has been amended for consistency within this chapter to conform to the ISO 10360 part 13 convention. As such, the error is notated as follows:

- $D_{\text{CC.99.7\%:CMV:O3D}}$, single-view distortion error.

8.5.4.2 Flat form distortion error

The assessment of the flat form distortion error determines the accuracy of flatness measurements throughout the measurement volume.

Dimensional requirements of material measures

In ISO 10360 part 13, a calibrated test plane is used to determine the flat form distortion error (i.e. material measure 3), with the standard noting that the manufacturer should specify the material and the surface of the material measure.

As per the standard, the length of the shorter side of the plane should be greater than 10% of the longest diagonal of the measurement volume L_0. The length of the longest side of the plane should be at least 50% of L_0.

It is acknowledged in the standard that the recommended dimensions for the test plane requirements cannot always be met, for example, due to cost or limited availability of flats suitable for very large measurement volumes. Therefore, there are allowances for the shorter side of the plane being smaller than recommended, but it should be noted that a larger test plane will allow for the flat form distortion error to be more clearly evaluated. If the requirement for the length of the longer side of the plane cannot be met, a travelling flat (a small test plane mounted on a travelling stage) or similar can be used.

Any reference plane supplied with the measurement system for system qualification purposes should not be used for this test to ensure independence.

VDI/VDE 2634 does not include a test of flat form distortion error.

Procedure

The test plane should be placed in six positions, as shown in figure 8.10. If these positions are not used, the modified procedure used must be reported. Global filters cannot be used when the MPE is assessed. It should be noted that additional MPE values can be defined by manufacturers, at their discretion, for special operating conditions, such as filters being applied to the measurement data by the measurement system.

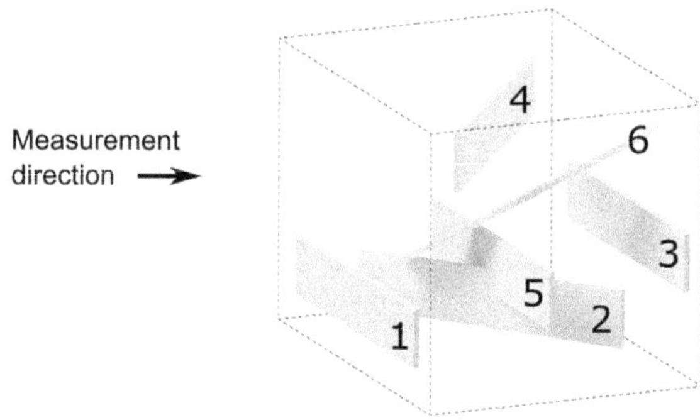

Figure 8.10. The six poses of the material measure in a cubic measurement volume.

Single-view measurements
For the single-view measurements, the material measure is measured in the six positions illustrated in figure 8.10. No repeat measurements are required in this case.

Multi-view measurements
For the multi-view measurements, ISO 10360 part 13 requires that the six positions be measured with no repeats, and the multiple views of the plane are registered and fused into one point cloud. Here, as in the test for volume distortion error, the data acquired for the six compulsory single-view measurements can be registered and fused into one point cloud in accordance with the manufacturer's guidelines.

Calculation of flatness errors
For each of the single-view flatness errors type (see below, for example $D_{\mathrm{Form.Pla.95\%:SMV.SV:O3D}}$), an error value is obtained for each of the six poses of the plane. For each of the multi-view flatness errors (see below, for example $D_{\mathrm{Form.Pla.95\%:SMV.MV:O3D}}$), a single error value is obtained for the registered and fused point clouds. For each of the 95% errors, the error is defined as the minimum distance between two parallel planes that contain 95% of the measured data. For each of the 100% (or 'All') errors, the error is defined as the minimum distance between two parallel planes that contain 100% of the measured data. If different fitting options exist in the manufacturer's software, ISO 10360 part 13 specifies that the option that minimises the maximum deviation from the ideal feature should be chosen (commonly referred to as Chebychev fitting). The following probing size errors are defined:

- $D_{\mathrm{Form.Pla.95\%:SMV.SV:O3D}}$, single-view flat form distortion error, 95% of data used;
- $D_{\mathrm{Form.Pla.95\%:SMV.MV:O3D}}$, multi-view flat form distortion error, 95% of data used;
- $D_{\mathrm{Form.Pla.All:SMV.SV:O3D}}$, single-view flat form distortion error, 100% of data used; and
- $D_{\mathrm{Form.Pla.All:SMV.MV:O3D}}$, multi-view flat form distortion error, 100% of data used.

8.5.5 Length measurement errors

The length measurement error test involves measuring the length measurement accuracy within the concatenated measurement volume, and per ISO 10360 part 13, is acquired using a ball beam comprising at least five independent sphere centre-to-centre measurements (i.e. material measure 4). There are three length measurement errors, the 'measurement volume length measurement error', acquired by making measurements of sphere centre-to-centre distances, using spheres defined by the full acquired point cloud.

Another (optional) error, namely the 'bidirectional' length measurement error, also exists in an annex of ISO 10360 part 13, to retain comparability with other parts of ISO 10360 (i.e. involving contact CMS measurements). In this case, a bidirectional measurement is a measurement that probes a distance from diametrically opposed directions potentially averaging out systematic measurement errors.

In earlier drafts of ISO 10360 part 13, a third error also existed, namely the 'unidirectional' length measurement error, which also existed to retain comparability with other parts of ISO 10360. To determine this error, representative points can be used (see section 8.5.1). In this case, a unidirectional measurement is one that probes a distance from multiple directions, potentially averaging out systematic measurement errors. At the time of publication of this book, it is unclear as to whether unidirectional length measurements will be included in the final version of ISO 10360 part 13, so to account for the eventuality where this section is returned to the document at some later date, the method is summarised here.

In the VDI/VDE document, the sphere spacing error measurement procedure is as described, but the data analysis process differs.

8.5.5.1 Dimensional requirement of material measures

In ISO 10360 part 13, the longest calibrated test length for each position should be at least 66% of the longest spatial diagonal of the concatenated measurement volume $L_{0:\mathrm{CMV}}$. The constraints on D_p are the same as for previous tests (see sections 8.5.3 and 8.5.4).

For measurements made per VDI/VDE 2634, more than one length must be used, because of the constraints placed on the test length by the specification. The constraints on D_p are the same as in ISO 10360 part 13, but in the VDI/VDE document the constraint for L_p is different, and in fact has two separate cases. When measuring VDI/VDE positions 1–3 (see table 8.2), the constraint is as follows:

$$L_{p1} \geqslant \frac{2}{3} L_{0:\mathrm{CMV}},$$

where $L_{0:\mathrm{CMV}}$ is the longest spatial diagonal of the concatenated measurement volume. When measuring VDI/VDE position 7 (see table 8.2), the constraint is as follows:

$$L_{p2} \geqslant \frac{2}{3} L_{\mathrm{min:CMV}},$$

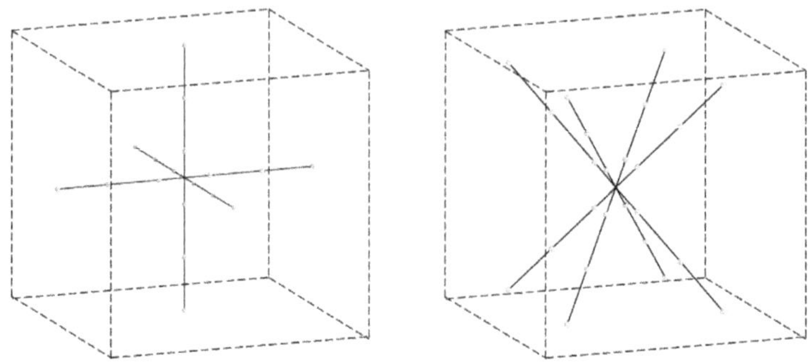

Figure 8.11. The seven poses of the ball beam. The measurement direction is arbitrary, since it is the concatenated measurement volume that is being tested in this case.

where $L_{\text{min:CMV}}$ is the length of the shortest side of the concatenated measurement volume. When measuring VDI/VDE positions 4–6 (see table 8.2), the longer distance of L_{p1} should be used if the whole feature fits within the measurement volume. Material measure 4 has a long sphere centre-to-centre distance that is suitable for L_{p1}, as well as a short sphere centre-to-centre distance that is suitable for L_{p2}.

8.5.5.2 Procedure for measurements of the concatenated volume

In ISO 10360 part 13, there are seven measurement positions varying in location and orientation (see figure 8.11). For each position, three repeat measurements are made and there are five measurement lengths on the specified length material measure, leading to 105 measurements in total. Four of the seven measurement positions are required to be the diagonals of the concatenated measurement volume. The remaining three positions can be chosen by the user, but it is recommended to align the final three positions parallel to the three co-ordinate axes. The user is free to choose the number and position of measurement views used to make these measurements.

In VDI/VDE 2634, measurements are also acquired in seven positions, but these positions equate to the seven VDI/VDE positions detailed in table 8.4 and figure 8.9. No repeat measurements should be made in this case.

8.5.5.3 Calculation of length measurement errors

A length measurement error can be obtained for each of the 105 measurements. These errors are calculated by fitting spheres to the measurement data of the ball beam. If different fitting options exist in the manufacturer's software, ISO 13060-13 specifies that the option that minimises the maximum deviation from the ideal feature should be chosen (commonly referred to as Chebychev fitting). The distances between the calculated centres of the spheres are the measured lengths. There are three categories of length measurement error, which are outlined here:

- $E_{\text{MVol:CMV:O3D}}$, measurement volume length measurement error in concatenated measurement volume.

In ISO 10360 part 13, a measurement volume length measurement error $E_{MVol:CMV:O3D}$ is obtained for each of the 105 measurements. The centre of each sphere is calculated by fitting a sphere to all the data acquired for each measured sphere. The error is calculated from the difference between the measured $L_{CC.meas}$ and calibrated $L_{CC.cal}$ values of the spheres' centre-to-centre distances. $L_{CC.cal}$ is equal to $L_{CC.Uni.cal}$:

$$E_{MVol:CMV:O3D} = L_{CC.meas} - L_{CC.cal}$$

- $E_{Bi:CMV:O3D}$, bidirectional length measurement error in concatenated measurement volume.

  In ISO 10360 part 13 Annex A, a bidirectional length measurement error $E_{Bi:CMV:O3D}$ can be obtained by supplementing the values for the length measurement error in the concatenated value with the values from the probing form dispersion error and the probing sizer errors. In the case where

  $$E_{MVol:CMV:O3D} + P_{Size.Sph.95\%:SMV.MV:O3D} > 0,$$

  $E_{Bi:CMV:O3D}$ has an upper limit which can be used to verify conformance:

  $$E_{Bi:CMV:O3D} < E_{MVol:CMV:O3D} + P_{Size.Sph.95\%:SMV.MV:O3D} + P_{Form.Sph.95\%:SMV.MV:O3D}.$$

  Then, in the case where

  $$E_{MVol:CMV:O3D} + P_{Size.Sph.95\%:SMV.MV:O3D} = 0,$$

  $E_{Bi:CMV:O3D}$ has upper and lower limits which can be used to verify conformance:

  $$E_{Bi:CMV:O3D} < P_{Form.Sph.95\%:SMV.MV:O3D} \text{ and } E_{Bi:CMV:O3D} > -P_{Form.Sph.95\%:SMV.MV:O3D}.$$

  Finally, in the case where

  $$E_{MVol:CMV:O3D} + P_{Size.Sph.95\%:SMV.MV:O3D} < 0,$$

  $E_{Bi:CMV:O3D}$ has a lower limit which can be used to verify conformance:

  $$E_{Bi:CMV:O3D} < E_{MVol:CMV:O3D} + P_{Size.Sph.95\%:SMV.MV:O3D} - P_{Form.Sph.95\%:SMV.MV:O3D}.$$

  In all cases, the probing form error from 95% of the data and the probing size error from 95% of the data are sampled and calculated once for the full set of 105 measurements.

- $E_{Uni:CMV:O3D}$, unidirectional length measurement error in concatenated measurement volume.

  In the earlier draft of ISO 10360 part 13 in which it was included, a unidirectional length measurement error $E_{Uni:CMV:O3D}$ was obtained with the use of representative points. Each sphere had its centre calculated by spacing representative points around the measured sphere and then fitting a sphere to the representative points (see section 8.5.1). The error was then calculated from the difference between the measured $L_{CC.Uni.meas}$ and calibrated $L_{CC.Uni.cal}$ values of the spheres' centre-to-centre distances,

$$E_{Uni:CMV:O3D} = L_{CC.Uni.meas} - L_{CC.Uni.cal}.$$

- $E_{Bi(VDI/VDE):CMV:O3D}$, VDI/VDE length measurement error.
  As in the current draft of ISO 10360 part 13, VDI/VDE 2364 does not allow for the use of representative points for the definition of errors. However, the calculation of $L_{CC.Bi.meas}$ using representative points, must be carried out as an initial step. Each measured distance $L_{CC.Bi.meas}$ is calculated by placing a single representative point on each sphere in the pair being measured, on opposing sides. The measured distance will, therefore, be the distance between the representative points, minus the calibrated radii of each sphere. Once these points are defined, the distances of the points to their respective sphere centres, $r_{meas:1}$ and $r_{meas:2}$, should be determined using the sphere centre locations calculated previously, when determining $L_{CC.Bi.meas}$. The errors on these measured radii, the difference from the calibrated values of $r_{cal:1}$ and $r_{cal:2}$, are then added onto the unidirectional length measurement error $E_{Uni:CMV:O3D}$. In VDI/VDE 2634, this error is referred to as length measurement error and notated as E, but has been amended for consistency within this chapter to conform to the ISO 10360 part 13 convention. As such, the error is notated as $E_{Bi(VDI/VDE):CMV:O3D}$, where

$$E_{Bi(VDI/VDE):CMV:O3D} = L_{CC.Bi.meas} - L_{CC.Bi.cal} + (r_{meas:1} - r_{cal:1}) + (r_{meas:2} - r_{cal:2}).$$

8.6 Compliance with specifications

In this section, the conditions in which a system can be considered to have had its performance verified are discussed, by reference to the formal definition of performance verification (see chapter 1). The conditions for passes, as well as 'hard' and 'soft' fails of the test are explained. The conditions for retesting following a failure are also detailed.

8.6.1 Formal definition of performance verification

The performance of the measurement system is verified if all the absolute values of the measurement errors are not greater than the MPEs as specified by the manufacturers, taking into account the measurement uncertainty as defined in ISO 14253 part 1 (ISO 2017). Therefore, the uncertainty in the calibrated material measure features must be included when comparing measurements with calibrated values. For example, an ambiguous result occurs when a recorded measurement error is larger than the relevant MPE, but smaller than the sum of the MPE and the calibration uncertainty. Rather than requiring repeated measurements to resolve the ambiguity, asymmetric burdens of proof are required. For a purchaser of a measurement system to prove their measurement system is performing outside of specifications, they must show that the measured MPE is greater than the stated MPE plus the relevant calibration uncertainty of the corresponding material measure. In contrast, for a manufacturer to prove that their system is performing as specified, they must show that the measured MPE is smaller than the stated MPE minus the relevant calibration uncertainty of the corresponding material measure.

The rule explained above is generally the case and represents a 'hard' fail case. However, in ISO 10360 part 13, there is an exception to this rule: in the case of the length measurement errors, a maximum of five of the thirty five sets of three repeated measurements can have one (and no more than one) of the three values of the length measurement outside the conformance zone (considered to be a 'soft' fail). Each set of such measurements with a single non-conforming value should then be remeasured three times in the appropriate position. If all three of the length errors for the repeated measurements are within the conformance zone, the performance of the optical CMS is verified for that measurement position. If more than five of the thirty five sets of three repeated measurements are outside the conformance zone, the test is failed in the same manner as any of the other tests.

8.6.2 Retesting after a failed compliance with specifications

If any of the measurement errors fall outside the MPEs, the CMS should be interrogated for any faults that may be causing failure and these faults corrected by the user. Following correction, reverification of the CMS is required as per the relevant standard/guidance document. However, if the second performance verification test is also failed, then manufacturer intervention is required before further measurements can be taken.

In VDI/VDE 2634, if the measurement error for a single measurement position on any test exceeds the specified MPE (accounting for measurement uncertainty as defined in ISO 14253 part 1 (ISO 2017)), then the measurements made at that position are repeated three times. If none of the new measurement errors exceed the stated MPEs, then the performance is verified. Otherwise, manufacturer intervention is required before further measurements can be taken.

8.7 A final note on the validity of performance verification

A final note should be made about the validity of these performance verification methods in testing whether an instrument is capable of performing a measurement task. As is stated clearly in the majority of performance verification documents, these procedures are to be explicitly used for testing whether an instrument meets a certain aspect of its specification and are often not applicable in a general case. For example, if a system is stated to have an MPE of $(5 + \frac{L}{100})$ μm for measurements made of sphere centre-to-centre distances (where L is in millimetres), the MPE for two spheres separated by 20 mm is 5.2 μm. However, this MPE is specific to the case described—using this example, it is not appropriate to assume that the MPE of a plane-to-plane distance of 20 mm is also 5.2 μm. Depending on the system used, the MPE for plane-to-plane measurements could be very different—it could be 5.02 μm,

205 μm or something else entirely. It is, therefore, highly important that the cases in which any one MPE is or is not relevant are noted, and that MPE values are not taken inappropriately out of context to make untrue assumptions about the performance of any one instrument. Fundamentally, performance verification always tests a system against some known reference, but real measurement cases often differ significantly from these often-idealised cases and so caution is required.

References

BIPM 2012 *International Vocabulary of Metrology—Basic and General Concepts and Associated Terms—JCGM 200* (Sèvres: Bureau International des Poids et Mesures)

Borges de Oliveira F, Bartscher M and Neuschaefer-Rube U 2015 Analysis of combined probing measurement error and length measurement error test for acceptance testing in dimensional computed tomography *Proc. Digital Industrial Radiology and Computed Tomography (Ghent, Belgium)*

Carmignato S, De Chiffre L, Bosse H, Leach R K, Balsamo A and Estler W T 2020 Dimensional artefacts to achieve metrological traceability in advanced manufacturing *Ann. CIRP* **69** 693–716

Drieschner R 1993 Chebyshev approximation to data by geometric elements *Numer. Algorithms* **5** 509–22

ISO 10360 2000 *Geometrical Product Specifications (GPS)—Acceptance and Reverification Tests for Coordinate Measuring Machines (CMM)* (Geneva: International Organization for Standardization)

ISO 10360 part 5 2010 *Geometrical Product Specifications (GPS)—Acceptance and Reverification Tests for Coordinate Measuring Machines (CMM)—Part 5: CMMs using single and multiple-stylus contacting probing systems* (Geneva: International Organization for Standardization)

ISO 10360 part 8 2013 *Geometrical Product Specifications (GPS)—Acceptance and Reverification Tests for Coordinate Measuring Machines (CMM)—Part 8: CMMs with optical distance sensors* (Geneva: International Organization for Standardization)

ISO 14253-1 part 1 2017 *Geometrical product specifications (GPS)—Inspection by measurement of workpieces and measuring equipment—Part 1: Decision rules for verifying conformity or nonconformity with specifications* (Geneva: International Organization for Standardization)

ISO/CD 10360 part 13 2020 *Geometrical Product Specifications (GPS)—Acceptance and Reverification Tests for Coordinate Measuring Machines (CMM)—Part 13: Optical 3D CMS* (Geneva: International Organization for Standardization)

VDI/VDE 2634 part 3 2008 *Optical 3D-Measuring Systems: Multiple View Systems Based on Area Scanning—Part 3* (Berlin: VDI/VDE)

VDI/VDE 2634 part 2 2012 *Optical 3D-Measuring Systems: Optical Systems Based on Area Scanning—Part 2* (Berlin: VDI/VDE)

Zhang S 2018 *High-Speed 3D Imaging with Digital Fringe Projection Techniques* (Boca Raton, FL: CRC Press)

www.ingramcontent.com/pod-product-compliance
Ingram Content Group UK Ltd.
Pitfield, Milton Keynes, MK11 3LW, UK
UKHW051856100926
8556IPUK00005B/645

9 780750 325226